THE SMOKE AND THE SPOILS

THE SMOKE AND THE SPOILS

Anti-Environmentalism and Class Struggle in the United States

JOHN HULTGREN

The MIT Press
Cambridge, Massachusetts
London, England

The MIT Press
Massachusetts Institute of Technology
77 Massachusetts Avenue, Cambridge, MA 02139
mitpress.mit.edu

The MIT Press would like to thank the anonymous peer reviewers who provided comments on drafts of this book. The generous work of academic experts is essential for establishing the authority and quality of our publications. We acknowledge with gratitude the contributions of these otherwise uncredited readers.

This book was set in Adobe Garamond and Berthold Akzidenz Grotesk by Westchester Publishing Services. Printed and bound in the United States of America.

Library of Congress Cataloging-in-Publication Data

Names: Hultgren, John, 1981- author.
Title: The smoke and the spoils : anti-environmentalism and class struggle
   in the United States / John Hultgren.
Description: Cambridge, Massachusetts : The MIT Press, [2025] | Includes bibliographical
   references and index.
Identifiers: LCCN 2024024742 (print) | LCCN 2024024743 (ebook) |
   ISBN 9780262552370 (paperback) | ISBN 9780262383233 (pdf) |
   ISBN 9780262383240 (epub)
Subjects: LCSH: Anti-environmentalism—United States—History. |
   Anti-environmentalism—Social aspects—United States.
Classification: LCC GE197 .H85 2024 (print) | LCC GE197 (ebook) |
   DDC 304.2/80973—dc23/eng/20241015
LC record available at https://lccn.loc.gov/2024024742
LC ebook record available at https://lccn.loc.gov/2024024743

10 9 8 7 6 5 4 3 2 1

EU product safety and compliance information contact is: mitp-eu-gpsr@mit.edu

Contents

Acknowledgments

Writing a book is a long and often solitary undertaking, but—since I began working on this project in 2016—I've been fortunate enough to find an intellectual community in many colleagues, friends, students, and activists.

At Bennington College, I've been welcomed into an institutional home that encourages me to ask big questions and to follow my inquiry wherever it may take me—including across disciplinary boundaries. My colleagues in the Environmental Studies and Society, Culture & Thought discipline groups, as well as in the Center for the Advancement of Public Action, have been nothing but supportive. Conversations with Lopa Banerjee, David Bond, Emily Mitchell-Eaton, Brian Michael Murphy, Kate Paarlberg-Kvam, Tim Schroeder, Susan Sgorbati, Paul Voice, Kerry Woods, and Marina Zurkow have all contributed to my thinking. My Bennington students have been patient enough to indulge me while I offered courses outside of my comfort zone; their eclectic insights and unyielding passion to change the world for the better are a vital source of inspiration.

I'm grateful for the continued friendship of my former colleagues at Northern Arizona University—especially Luis Fernandez, Paul Lenze, Sean Parson, and Brian Petersen—who, among other things, included me in their always lively, thought-provoking reading group when COVID-19 forced our social interactions online. For several years, I had the good fortune of engaging in a mutual dialogue with Williams College colleagues through our political theory writing group. Greta Snyder, Laura Ephraim, and Will Stahl read over early iterations of several chapters and offered incredibly constructive feedback. Dimitris Stevis—my mentor, collaborator, and friend—not only provided comments on several draft chapters, but continues to serve as a model for rigorous scholarship grounded in a commitment to praxis.

Several chapters of this book were first presented at annual meetings of the Western Political Science Association, where I received incisive suggestions from members of the always supportive Environmental Political Theory community. Over the years, I've benefited enormously from the feedback and collegiality of Andrew Biro, Peter Cannavò, Teena Gabrielson, Anatoli Ignatov, Robert Kirsch, Gregory Koutnik, Jenn Lawrence, Brian Lovato, John Meyer, Ross Mittiga, Sean Parson, Em Ray, Cedar Welker, Harlan Wilson, Mary Witlacil, Steve Vogel, and Rafi Youatt (as well as numerous others who I'm sure I'm unintentionally leaving out).

At the MIT Press, Beth Clevenger, Anthony Zannino, and the meticulous copy editing team have been a pleasure to work with. The three anonymous reviewers offered comments that were both generous and generative. A number of environmental activists graciously took the time to chat with me: Alex Amend, Anahkwet (Guy Reiter), Dale Burie, Veronica Coptis, Ashley Funk, Al Gedicks, Karan Ireland, Richard Kazis, Justinn Overton, Ramsey Sprague, Stephen Stetson, and Dustin White. Whether I ended up quoting them directly or not, their insights have enriched my thinking.

I've been working on this book for almost nine years, but the roots of my interest in the topic go back to my childhood in Menominee, Michigan—where I spent summers swimming in the bay with smokestacks from a paper mill perched in the distance; where the mills, chemical plant, and foundries have provided much-needed jobs but at the cost of polluted air and waters; and where the population has steadily declined and politics have lurched dramatically rightward over the past several decades. My interest in the *oikos* is informed by this place, and all of my extended family—grandparents, aunts, uncles, cousins, and friends—who helped make it home.

In this vein, I'm grateful for the love and support of my Mom and Dad, Julia, Max, Laura, Ken, Olivia and Eva, Jerry, Kathy, Tim, Ingrid, Henrik, Paige, and Flor. I truly could not have written this without the kindness and humor of my wife, Mary, and the wonder and excitement provided by my kids Emmaline and Charlie. Watching the joy on their faces as they explore the woods, splash in the water, or run through the yard is a wonderful reminder that there's a world worth fighting for.

INTRODUCTION: PUTTING ANTI-ENVIRONMENTALISM TO WORK

The Shell Pennsylvania Petrochemicals Complex is a dense network of manu-facturing plants, bridges, freight cars, cooling towers, storage tanks, genera-tors, and combustion turbines that sprawls across 386 acres on the banks of the Ohio River in Beaver County, Pennsylvania. When fully operational, the six-billion-dollar facility is expected to employ roughly six hundred people, produce 1.6 billion metric tons of plastics, and emit 2.2 million tons of carbon dioxide per year (in addition to volatile organic compounds, nitrogen oxides, sulfur dioxide, carbon monoxide, and particulate matter) (Clean Air Council 2014). On August 13, 2019, something else was hanging in the air, though. At a formal White House event-turned-political-rally, former president Donald Trump—flanked by construction workers in yellow vests—lofted up an appeal to unionized labor:

> I'm truly honored to be here with the amazing energy workers and construction workers, . . . the craft workers who make America run and who make America proud. . . . Today we celebrate the revolution in American energy that is help-ing make our economy the envy of the world. This shell petrochemical plant in Beaver County, Pennsylvania . . . one of the single biggest construction projects in the nation, and made possible . . . by clean, affordable, all-American natural gas. (ABC News 2019)

Linking energy production with the well-being of the American working class, Trump's economic nationalism tethered to expanded fossil-fuel production struck the right chords with those present. "We're not only unleashing Ameri-can energy," he said, to cheers from the audience, "we're restoring the glory of American manufacturing and reclaiming our noble heritage as a nation of builders again" (ABC News 2019).

It soon came to light, however, that at least some of the workers in attendance were not there by choice. Journalists reported that "prior to Trump's appearance, a contractor for Shell sent a memo to union leaders instructing members whose attendance was not recorded by time clock scans would not receive overtime pay for the week." The memo also warned that "no yelling, shouting, protesting or anything viewed as resistance [would] be tolerated" (Hiltzik 2019; see also Litvak 2019). In a post–Citizens United world, oil and gas production wasn't all that had been unleashed: Shell found itself newly free to engage in election-related communication with its employees, even at the expense of the workers' time or their jobs (Harvard Law Review 2014). One union worker noted that the wages and benefits lost by refusing to attend the speech could amount to $700.

For Shell, the project is part of a broader plan to expand into plastics production, a transitional strategy at a moment where the future of oil and gas seems increasingly imperiled. For many in the Ohio River Valley, the complex is also appealing: in a region decimated by deindustrialization and shaken by the empty promises of the fracking boom, plastics offer a potential pathway to economic stability (O'Leary 2021). The Shell "cracker plant" is part of a vision for an Appalachian storage hub that includes "more cracker plants, methanol facilities, underground storage facilities . . . plastics manufacturing facilities and fractionators [and] . . . a conjoining infrastructure that will run down the Ohio River . . . about 500 square miles" (White 2021). The allure of the coming "petrochemical renaissance" has captured support from Republicans as well as many centrist Democrats (American Chemistry Council 2017; US Senate 2019). The state of Pennsylvania was so desperate to attract the plant that they granted Shell a sweetheart deal, $66 million dollars in annual tax credits, multimillion-dollar state grants, and additional tax waivers and abatements spanning twenty years into the future (Perisco, Alternburg, and Simeone 2021). One Beaver County commissioner saw it as the only realistic option for the area: "If we aren't making things for a living, if people aren't using our hands and their minds and we're just trying to be a service economy, I don't see how we survive in that kind of an environment" (Frazier 2022).

But other locals worry about their loved ones surviving in an environment contaminated by the pollution that will be emitted from the plant. Some who lived in close proximity decided to move away, fearing its impacts on air quality. As one resident put it, "We looked at . . . maps projecting particularly risky areas with air quality—the elementary school is rated right in the high-risk area. . . . Our children are going to be spending 8 hours a day in a high-risk area" (Frazier 2022). In one health impact study, the Clean Air Council underscored that the pollutants emanating from the facility, including NO_x, SO_2, PM2.5, and ozone, "have been linked to increased risk of respiratory and cardiovascular symptoms, and with increased mortality rates" (Clean Air Council 2014, 30). The Sierra Club adds that the Appalachian petrochemical hub requires the construction of six new pipelines whose location on the river threatens "the source of drinking water for 5 million people" (Sierra Club nd). For Dustin White, a native of West Virginia coal country and a long-time environmental activist in the Ohio Valley region, this vision of economic flourishing through petrochemicals adds present insult to historical injury. "We already have a long history of being contaminated by the petrochemical industry," he explained to me, noting the prevalence of the toxic chemical C8 (or perfluorooctanoic acid) in the region: "Here we now finally have the coal industry going belly up, and we're gonna have to deal with the legacy costs of that, and they're wanting to throw another toxic industry right on top of it" (White 2021).

The debate over the petrochemical complex is, in many respects, a microcosm of the broader national struggle over how to best provide people with two vital material needs: (1) well-paying, dignified jobs and (2) a safe and stable environment. The dominant framing of this struggle is one that pits workers, on the one side, seeking steady jobs and good wages, against environmentalists, on the other, fighting for clean air, water, and climate mitigation. Over the past fifty years of US politics, industry's most powerful counterargument against environmental protections has been that policies intended to safeguard the environment impose undue harms on ordinary working-class people. For those seeking to prolong fossil fuel use, ward off climate mitigation, and scale back measures aiming to clean up our air and water, such a framing has been remarkably successful. To an even greater extent than the faux-scientific strategies of climate denial advanced by the likes of Republican politicians and the Heartland Institute, this zero-sum dichotomy between jobs and environmental protection, workers and environmentalists, lies at the foundation of the US anti-environmental movement.

But with every passing heat wave, wildfire, and sunny-day flood, the sweeping effects of today's environmental crises become more apparent. For forty-seven consecutive years, global temperatures have surpassed the twentieth-century average. Heat-related deaths have risen by two-thirds over the past twenty years. And the more quotidian air pollution caused by the burning of fossil fuels contributes to over a million deaths every year (Romanello et al., 2022). In 2023, the warmest year on record, oil companies reaped record profits, while wildfires raged across the world's boreal forests. Drought intensified food insecurity and fueled mass migration out of Central America, while flooding displaced several million across the Horn of Africa. Glaciers "suffered the largest loss on record," and sea level rise continued to accelerate (World Meteorological Organization 2024). The sixth assessment report from the Intergovernmental Panel on Climate Change (IPCC) concludes with the ominous warning that "any further delay in concerted anticipatory global action on adaptation and mitigation will miss a brief and rapidly closing window of opportunity to secure a livable and sustainable future for all" (IPCC 2022, 33).

Clearly, the time for real, meaningful action is now. Why, then, aren't we taking such action? What social forces are standing in the way? And to what sorts of strategies and political visions should those of us concerned for the future turn? This book attempts to shed light on these questions, and in doing so to sketch out a political theory of anti-environmentalism that is grounded in what too many such analyses parcel off as a separate locus of study: the realities of class struggle.

THEORIZING CLASS

Across the political spectrum today, the environment—as both a material sphere of life and a socially resonant symbol—is embedded in a renewed reckoning with class. For conservatives, extractive laborers hold the key to making America great again, while environmental elitists, out of touch with the needs of the heartland, stand in the way. For liberals, the professional class, comprising well-educated suburbanites and technocratic experts, will solve the crisis of climate change through public/private partnerships, state subsidization of eco-entrepreneurs, and the creation of carbon markets. For leftists, efforts to build a Green New Deal lie at the foundation of attempts to reinvigorate working-class consciousness and revolutionize US politics. Examining how the environment and class have been catapulted to such prominent positions within US politics requires grappling with environmental dimensions of political struggles that have long existed but remain underacknowledged.

To understand the environmental dimensions of class struggle, it is necessary to first explain what we mean by class. In popular commentaries, class is often reduced to an identity defined by one's occupation, education level, or income. These metrics all capture important aspects of our day-to-day lives, but the analyses that rely on them tend to obscure the fact that class is, at its core, a social relation. Class is determined by our relationships to labor and the process of production: capitalists control the means of production, and workers do not; capitalists buy labor power, and workers are forced to sell theirs. As a consequence, the material economic interests of capitalists—who seek to extract as much value as possible from workers in order to reinvest in new forces of production and thereby lower costs of production—are antagonistic to those of laborers, who seek better wages and working conditions. The objective location of economic actors in relation to the means of production is referred to as *class-in-itself*. The *working class-in-itself* comprises those who have nothing to sell but their labor; the *ruling class-in-itself* comprises those who buy labor power to propel the means of production that they own.

And yet, even during the golden age of industrial capitalism, workers did not always think of themselves as "working class" and often did not act on the basis of any working-class interests; indeed, workers' understanding of their place in society and their relation to other social actors and forces has varied dramatically across time and place. "Class," as E. P. Thompson notes, "is defined by men as they live their own history" (1963, 11). This sociocultural dimension of class analysis—the presence (or absence) of class-based social organizations, a shared political consciousness, and a sense of solidarity—is referred to as *class-for-itself*. The working class has historically mobilized as a class-for-itself through "workingman's" organizations, labor unions, and social-ist parties, wherein workers have shared grievances and built bonds of mutual support, trust, and friendship through participation in organizing campaigns and strikes, educational programs, and recreational activities.

In contrast to the working class, which must speak for itself in order to be heard, the ruling class-in-itself is typically careful not to appear as a class-for-itself—preferring to conduct its rites of solidarity behind closed doors while letting politicians, technocratic experts, and popular commentators proclaim its service to the greater good in public fora. The ruling class is rarely uniform in its politics (there exist major "fractions of capital," to use Nicos Poulantzas's phrase), but, in the United States, large segments of capital have historically organized and coordinated their interests through membership in trade groups (e.g., the National Association of Manufacturers and American Chemistry Council), business networks (e.g., the Chamber of Commerce), political advocacy organizations (like the Liberty League of the 1930s or the American Legislative Exchange Council of today), and political parties. More-over, during moments of social upheaval and instability, the ruling class often acts as a relatively unified bloc by cutting wages, laying off workers, moving capital abroad, funneling money to right-wing think tanks and politicians, and pressing for particular types of state interventions: corporate bailouts; deregulation in the areas of finance, labor, and the environment; and, in some cases, support for moderate reforms that work to stabilize social relations and enable a return to business as usual.

This distinction between class-in-itself and class-for-itself is strategically indispensable insofar as it allows us to identify, in broad strokes, those who benefit materially from the existing political economic order and those who suffer from it; those who we should stand alongside, and those against whom we must stand opposed. These objective interests (class-in-itself) do not lead in any straightforward way to identities and collective action (class-for-itself); such is the stuff of political struggle. But without the strategic and moral clarity that comes from delineating who wins and who loses from the existing political economic order, we risk being led astray by popular tropes of reactionary workers and median voter logics that lavish attention on the upper-middle-class, suburban demographic at the expense of the actually exploited. At the same time, it is worth pointing out several deficiencies in the aforementioned orthodox framework and revising our conception of class accordingly. In the discussion that follows, I briefly situate my approach to class within four debates, related to (1) the impacts that shifting forms of production have had on class composition; (2) the ways in which racism has affected the process of class formation; (3) the relationship between "productive" and "social reproductive" labor; and (4) the role that the state plays in mediating class relations. I then pivot back to environmental politics and offer a preliminary overview of how the politics of nature have factored into class struggle in the United States.

First, the dichotomous distinction between the ruling class and working class is both a normative call to arms and an ideal-typical description of a future to come (assuming the continued extraction of surplus value,[1] on one side, and the presence of political mobilization, on the other). It has always been an imperfect approximation for the actually existing positions that many people occupy in relation to production but has grown increasingly fraught over the past century. By the mid-twentieth century, the working class had come to include "secretaries and executives, nurses and corporate lawyers, teachers and policemen, computer operators and executive directors, . . . all separated from the means of production and compelled to sell their labor power for a wage" (Przeworski 1977, 356). This historical moment, marked by the ascendance of a "middle class" and the appearance of automation and deindustrialization (already producing long-term involuntary unemployment in former industrial strongholds), was one in which material reality not only demanded that any mechanistic relationship between the working class and emancipatory change be consigned to the ash heap of history but also brought to the fore—via the global political explosions of the 1960s—the distinct possibility that the revolutionary subject would not be found in the industrial factory or field. Such shifts in class composition made it more difficult to mobilize a working class-for-itself; the oft-invoked "middle class" was, as many recognized, generally averse to thinking of itself as working class (see, e.g., Miliband 1969, Mills 2002) and unions, as autoworker and scholar-activist James Boggs regretfully observed, had made a conscious choice not to struggle in solidarity with the growing ranks of the unemployed (1963, chap. 1).

Second, and with this in mind, any discussion of the working class needs to consider how this form of collective agency has historically been shaped by the realities of racism. In historical fact, racism has always been integral to capital accumulation. In his seminal work, *Black Marxism* (1983), Cedric Robinson argued that social divisions inherited from feudalism were part and parcel of the process of class formation in early capitalist Europe. The power of national and racial identities quickly foreclosed the possibility of a united working class emerging out of shifts in relations of production. In the US context, white workers have frequently chosen to join in racial solidarity with white elites rather than in class solidarity with workers of color (Du Bois 1992; see also Saxton 1990; Roediger 1991; Ignatiev 1995; Nelson 1996; Foley 1997; Brodkin 1998; Lipsitz 1998; Hale 1999; Mills 2004; Olson 2004). The working class-in-itself has always been multiracial, multinational, and multi-gendered, but the organizational expressions of the working class-for-itself were, until the latter half of the twentieth century, generally fragmented along racial lines, in particular (with some notable exceptions that I discuss in the chapters to come). As a consequence, there exists a long-standing disjuncture between the dominant image of the working class (comprising white male factory and extractive-sector workers) and the material constitution of the working class.[2] These discursive and organizational exclusions from the working class have, of course, served the interests of capital; the devaluation of racial and ethnic minorities has provided a cheap pool of labor for capital to exploit and has divided working-class movements, sapping their power in the process.

Third, accompanying this racial exclusion is a gendered division that stems from the sharp boundary that is typically drawn around the exploitation experienced at the site of production and that associated with the myriad noncommodified forms of production on which capital accumulation relies. The orthodox logic here is not entirely without merit: the site of production—whether a factory, field, mine, retail store, or restaurant—is where value is produced, where exploitation is experienced, and where organized resistance can most directly block the advance of capital accumulation. As socialist feminists remind us, however, wage labor and capital accumulation could not "exist in the absence of housework, child-raising, schooling, affective care, and a host of other activities that serve to produce new generations of workers and replenish existing ones, as well as to maintain social bonds and shared understandings" (Fraser 2014; see also Federici 1974; Hartmann 1979; Bhattacharya 2017). Class-in-itself is structured not only through relations of production but also through relations of *social reproduction*, "the everyday activities of maintaining life and reproducing the next generation" (Bakker 2007). In the realm of social reproduction, the drive for accumulation often brushes up against many of the requirements and desires of life—leisure time, raising and educating children, caring for the sick and elderly, and having ample food to eat, clean air to breathe, and potable water to drink.

This separation between the home, a site of subsistence, and the workplace, a site of production—gradually institutionalized over the course of the late eighteenth through mid-nineteenth centuries as distinctly *gendered* sites—obfuscates this interdependence and severs the connections between movements emanating from material insecurity experienced in the home, neighborhood, and local community and those emanating from the experience of exploitation in the workplace. The multiple forms of marginalization and exclusion that, say, housewives of the 1950s experienced are not reducible to capitalist exploitation, but the financial precarity, industrial pollution, and everyday stresses caused by long working hours and stagnant wages echoed into the home in ways that we need to examine. The struggles in the home and workplace are not one and the same, but they are interconnected through the intermediary of labor. To suggest that a stay-at-home parent in a working-class neighborhood (or, for that matter, someone who was laid off from a factory and begrudgingly took up work in the informal economy) is not "working class" because their labor is outside of the formal circuits of accumulation conceals the true breadth of capitalist exploitation and immiseration; erases the crucial bridges that connect the workplace to the local community, neighborhood, and home; and ignores the commonsensical understandings of class that propel people into action.

Fourth, beyond the labor of social reproduction, another necessary condition for capital accumulation is state power (Polanyi 1944; Fraser 2014). In addition to its "monopoly on the legitimate use of physical force" (Weber 1919), the state is an institutional form that shapes and constrains the field of class relations—so much so that "it makes no sense whatever to speak of a social field of class division of labour and class power existing prior to the state" (Poulantzas 1980, 39). Contra the orthodox position, the state is not reducible to a tool of capital—"a committee for managing the common affairs of the bourgeoisie," as Marx and Engels famously put it in the *Manifesto*; rather, it is the primary extra-economic terrain of class struggle, albeit one in which the deck is stacked against the working class. On the one hand, capitalism relies on the state, not only to enforce property rights and contracts or to respond violently to working-class dissent but also to steady social and ecological relations amid the tumult caused by the commodification of land, labor, and money (Polanyi 1944). On the other hand, however, "it is precisely because capitalism cannot secure through market forces alone all the conditions needed for its own reproduction that it cannot exercise any sort of economic determination in the last instance over the rest of the social formation" (Jessop 2002, 11). Capital's reliance on a democratic state opens up space for working-class resistance. But what political opportunities does the relative autonomy of the state present to the working class?

In his 1977 essay, "The Ruling Class Does Not Rule," sociologist Fred Block offers insight into this question. In explaining capitalism's stubborn persistence, Block asserts that "the capacity of capitalism to rationalize itself is the outcome of a conflict among three agents—the capitalist class, the managers of the state apparatus, and the working-class" (Block 1987, 52). The capitalist (or ruling) class, though fragmented, tends to focus on short-term profits and support policies that would, if adopted, lead to structural crises. However, state managers, "those at the peak of the executive and legislative and branches of the state apparatus" (1987, 201), have an incentive to ensure the smooth and stable functioning of the political system. There are many historical cases—for example, the New Deal of the 1930s, the "command and control" environmental laws of the late 1960s and early 1970s—during which, at moments of social upheaval or economic instability, the state has enacted regulations contrary to the interests of particular capitalists (arguably even contrary to the short-term interests of capital as a whole). State managers are nonetheless predisposed to create the conditions for continued accumulation, both because the state's coffers depend on revenues derived from corporate activity and because the state's legitimacy is reliant on continued business confidence (and the economic growth and jobs that accompany this confidence). These are structural mechanisms—beyond the interpersonal and interorganizational ties that capitalists have with politicians and regulatory agencies, beyond the ideological pull of free-market logics, and beyond the class position of state managers themselves—through which capital constrains the decisions and practices of the state. Working-class demands may be met (usually partially), but the resulting regulatory structure is often translated into the stabilization of accumulation, and thus the long-term interests of capital in general.

And yet, Block's analysis also suggests that there are historical conjunctures during which these structural constraints deteriorate, albeit temporarily:

> There are certain periods—during wartime, major depressions, and periods of postwar reconstruction—in which the decline of business confidence as a veto on government policies doesn't work. These are periods in which dramatic increases in the state's role have occurred. (1987, 66)

Such crises create punctuated equilibria (Baumgartner and Jones 1991) in which the conditions of possibility for working-class movements to mobilize state power toward their desired ends increase. Insofar as the state remains the only political institution powerful enough to provide a meaningful check on transnational capital, attention to these conjunctural crises—a global pandemic, for instance—are of the utmost importance if leftist movements hope to institutionalize any of their policy goals, like a Green New Deal.

The particular form of the state also matters, though. The unique institutional structure of the US government poses a number of additional barriers to working-class power: (1) the rules of elections and the structure of electoral districts—that is, first-past-the-post elections and single-member districts—almost inevitably produce a two-party system with which working-class movements must contend; (2) both the Senate and Electoral College provide disproportionate power to smaller states, which has amplified the power of the conservative Republican Party (with its anti-labor and anti-environmental platform) in recent years; (3) due to the peculiarities of the contemporary filibuster, the US Senate has effectively become an institution wherein a supermajority is required to pass anything—a phenomenon that has continually squelched progressive legislation; (4) electoral redistricting has become hyper-politicized, which the Republican Party, in particular, has used to gerrymander state and national legislative districts to its advantage; (5) the constitutional authority granted to private property, long codified through the notion of "corporate personhood," is interpreted by an increasingly conservative federal judiciary with lifetime appointments; and (6) the managers of the state apparatus are not career bureaucrats but politicians and political appointees whose fealty is to the executive who appointed them, the interests of the corporations with whom they'll almost certainly work after their foray into government, and/or the ideology of the party base on whom their political future rests. In theory, *some* of these institutional characteristics—lifetime appointments for federal judges, political appointments, use of the filibuster to block legislation, and so on—could be used to the advantage of the Left, but the current balance of power is tilted dramatically rightward. This is no mere historical contingency, but reflects the "cumulative impact of the series of historic defeats suffered by the American working-class" (Davis 1986, 7).

THE ENVIRONMENT AND CLASS STRUGGLE

Each of these aforementioned debates about class—regarding the shifting contours of production, racism, social reproductive labor, and the state—has vital implications for environmental political struggles. To begin to flesh out these linkages, I now turn to a fifth debate: how the politics of nature factors into class relations. Raj Patel and Jason Moore argue that "Nature" and "Society" emerged as distinct domains of life alongside colonialism in the fifteenth and sixteenth centuries:

> Where European capitalism thrived was in its capacity to turn nature into something productive and to transform that productivity into wealth. This capacity depended on a peculiar blend of force, commerce, and technology, but also something else—an intellectual revolution underwritten by a new idea: Nature as the opposite of Society. (2017, 46)

The colonial ecological revolution, as Carolyn Merchant terms it, transformed "a society in which humans, animals, plants, and rocks were equal subjects . . .

to one dominated by transcendent vision in which human subjects were separate from resource objects" (1987, 273). This emergent nature functioned as "a way of organizing life" that systematically devalued both its nonhuman inhabitants and those human populations deemed to be within its purview. The colonial organization of life provided ideological legitimation for a transformation that was simultaneously ecological and social—land and resource grabs enriched and emboldened colonial powers, in the process clearing the way for private property and, eventually, entrenching capitalist relations of production. As industrial capitalism emerged, nature was catapulted into a crucial site of class struggle, and class relations began to impact an emerging set of concepts that we now think of as part and parcel of environmental politics—wilderness, ecology, and conservation.

The launching point for my analysis is the notion that capital accumulation—the process by which ever-expanding profits accrue in the hands of those who own the means of production—is, as Alf Hornborg puts it, "an ecological process" (1998). It is also a violent process—red in tooth and claw, and covered in soot. As Rosa Luxemburg realized,

> Accumulation, with its spasmodic expansion, can no more wait for and be content with a natural internal disintegration of non-capitalist formations and their transition to commodity economy, than it can wait for and be content with the natural increase of the working population. Force is the only solution. (1913, 351)

The use of force that enabled accumulation initially had three components: the forceful removal of peasants and indigenous populations from the commons; the forceful removal of enslaved African peoples from their homelands (and their subsequent forced labor in the Americas); and the forceful removal of "resources" from the land, forests, and waters. These resources—the metaphorical fruits of nature's bounty—were not shared equitably; rather, the enclosure and appropriation of the commons predictably worsened the lot of *commoners* who were rapidly forced into wage labor, and had even more dire impacts on indigenous populations. The rise of industrial capitalism was thus contingent on an unprecedented capture of nonhuman flows and their transfer into the hands of private owners, particularly those invested in mechanized production. Marx and Engels argued,

> Subjection of Nature's forces to man, machinery, application of chemistry to industry and agriculture, steam-navigation, railways, electric telegraphs, clearing of whole continents for cultivation, canalization of rivers, whole populations conjured out of the ground—what earlier century had even a presentiment that such productive forces slumbered in the lap of social labour. (1992, 8)

But "social labor" alone can't explain the "astonishing accumulation of capital" that occurred in the transformation of the wild frontier into early industrial centers. Natural processes—fertile soils, roaring rivers, lush forests—that developed over millennia were a precondition for the production of any value (see, e.g., Cronon 1992, 149; White 1995). These ecological processes were increasingly disrupted by the forms of production emerging in the nineteenth century—the shift to steam power and factories powered by combustion engines, as well as industrial agriculture relying on chemical fertilizers. John Bellamy Foster and his collaborators have referred to this as a *metabolic rift*: "a rift in the metabolic exchange between humanity and nature," first theorized by Marx in the context of "the robbing of the soil of the countryside of nutrients and the sending of these nutrients to the cities in the form of food and fiber, where they ended up contributing to pollution" (Foster, Clark, and York 2010, 45).

The mid-nineteenth-century ascendance of what Andreas Malm terms "fossil capitalism"—the conjuncture at which fossil fuels become "the general lever for surplus value production" (2013, 51)—put this whole process on a steadily increasing regimen of steroids that irrevocably transformed not only the pace and scale of production and entire ecological systems but, eventually, *geological* processes themselves. The use of fossil fuels—in which the "time and space required for photosynthesis hundreds of millions of years ago" are compressed into specific place-bound reservoirs (Malm 2013, 56)—paradoxically freed capital from the social and environmental limitations of being situated in a particular place and moment in time: factories could be moved away from energy sources, like rivers and forests, where a labor force needed to be imported and a whole infrastructure for communal life had to be built, and into the cities, where a vast population of surplus labor existed (these factories were, of course, propelled not only by fossil fuels but by primary products, like cotton, that entered into commodity form through the coerced labor of enslaved peoples and the dispossession of indigenous peoples) (Jobson 2021). As a general rule, the power of capital has intensified as the burning of fossil fuels—and environmental degradation at a geological scale—has increased (Malm 2016, 314–316).

And yet, capitalism is built on a structural foundation rife with fault lines, none bigger than the fissure separating workers and capitalists. Capitalists are in continual pursuit of what Patel and Moore term "Cheap Natures": "a rising stream of low-cost food, labor-power, energy, and raw materials to the factory gates" (2017, 53). Wage workers' subordinate position within this structure ensures that they also require cheap natures—that is, access to food, water, energy, and sources of sustenance that they can afford with their often meager wages. There is a contradiction between the economic *and ecological* require-ments of the working class and those of the ruling class—the exploitation of workers at the site of production is woven into the material hardships that they face at home, in the form of financial precarity and pollution, and is irrevocably connected to the decimation of the homes and habitats of nonhuman species.

With this in mind, I define the working class-in-itself as those who are compelled to sell their labor and, as a result, experience *exploitation at work and/or material insecurity at home*. This definition encompasses the productive and social reproductive labor that are both so central to capital accumulation, as well as the economic and environmental degradation that such workers experience in their day-to-day lives. The exploitation, predicated on the extrac-tion of the value that workers produce, manifests itself in the asymmetry of the wage relation, while the material insecurity arises in the form of precarious living conditions, continual exposure to polluted air or water, an inability to afford ample or decent food, and/or routinized harassment at the hands of local law enforcement. In addition to wage laborers, this approach to the working class includes people like unemployed single mothers getting by on Medicaid and SNAP benefits and elderly couples subsisting on their social security while health care and housing costs mount. My definition is not so expansive as to encompass those experiencing any form of oppression or hardship, but it is broad enough to capture multiple dimensions of contemporary inequality and to more accurately reflect commonsensical understandings of class, as people live their own history.

I define the ruling class-in-itself as comprising those who own the means of production, who purchase labor power and thereby enact exploitation, and whose interests are *necessarily* anti-environmental. Individual capitalists may well be—and, historically, have often been—ardent preservationists and environmental champions in their individual lives (Farrell 2021), but they operate within a market structure in which the drive to lower costs of production is a logical necessity insofar as it heightens economic competitiveness and thereby brings in higher profits. Movements for the preservation of "wild nature," for example, emerged in the nineteenth century alongside industrialization and were propelled by political and economic elites who, as Raymond Williams so aptly put it, "change their clothes at weekends or when they can get down to the country; join appeals and campaigns to keep one last bit of England green and unspoilt; and then go back, spiritually refreshed, to invest in the smoke and the spoil" (Williams 1972, 59). There are, of course, companies who genuinely attempt to "green" their operations in order to occupy niche markets, but such actors remain constrained by the strictures of accumulation; for larger corporations, any gains in resource efficiency that come from adopting green technologies in production processes are offset by a simultaneous expansion in the volume of production (and thus the need for more and more raw materials); for smaller companies, the ability to maintain a sustainable business model exists in inverse proportion to the popularity of the product (with greater profits bringing more competition and a necessary drive to lower production costs). It is possible that so-called green capitalists—for example, in the renewable energy sector—will play a key role in the transition away from fossil fuels, but they are structurally locked in to an exploitative relationship with their workers and with nature. The eco-capitalist is a contradiction in terms.

This is particularly apparent with the most powerful fraction of today's ruling class: fossil capital. Malm and the Zetkin Collective (2021) note that fossil capital comprises two segments: *general fossil capital*—that is, "capital for which fossil fuels are a necessary auxiliary in the production of other commodities"; and *primary fossil capital*—that is, the fossil fuel industry. For general fossil capital—from auto manufacturers to Silicon Valley start-ups and from pharmaceutical companies to private equity funds—policies proposing to cut emissions and develop alternative non-fossil-fuel energy sources present a major challenge that would require deep changes to their operating practices; for primary fossil capital, the stakes are even higher: oil and gas, coal, and the petrochemical industries face "an existential crisis, because the prevention of dangerous anthropogenic interference with the climate system ultimately requires that [they cease] to exist" (Malm and the Zetkin Collective 2021, 17). Over the past century and a half, the standard bearers of fossil capital have not only wreaked havoc on the environment but have embraced and actively advanced many of the most reprehensible ideologies and movements in US politics: from Standard Oil's violent strikebreaking to DuPont's funding of the Liberty League, and from the Koch family's dalliance with the John Birch Society to Chevron and Marathon Petroleum's continued donations to Republican members of Congress who refuse to speak out against the Capitol insurrection (Kahn and Mehrotra 2021; Kroll 2021). This is to say nothing of the violent neocolonial adventures that have been led by extractive interests, both on Native American lands and abroad—Texaco in Ecuador, Shell in Nigeria, Total and Chevron in Burma (the list goes on). Primary fossil capital has functioned as the vanguard of the ruling class, but instead of leading the masses to the barricades they've chauffeured the elites to their gated communities and guarded islands while preparing carefully sculpted talking points that link the future of fossil fuels to that of the working class.

In sum, class relations in the United States have been, and continue to be, structured through environmental politics, broadly conceived to include material struggles over the control of actually existing nature (forests, fields, rivers, lakes, oil, coal, minerals, animals, etc.) and ideological struggles to authoritatively define nature (ecology, wilderness, the environment, sustainability, etc.). For analytical purposes, we can start by identifying four environmental dimensions that are virtually always present in class struggles: (1) the ruling class-in-itself requires authority and control over the actually existing nonhuman lives, forces, and flows (water, trees, minerals, coal, oil and gas, etc.) that are necessary for production; (2) as production has increased, the demand for non-human resources has proceeded apace, and the efforts of capitalists to control ever-increasing swathes of nature have necessarily expanded; (3) the resulting struggles to control, access, and enjoy the fruits of nature have routinely produced organizations and forms of solidarity that have galvanized a sense of class-for-itself among ruling and working classes alike; and (4) environmental concepts (nature, wilderness, ecology, environment, sustainability, etc.) have played crucial ideological roles in both legitimating and challenging dominant class relations, as well as in mediating the relationships among class and other vectors of social life, including race, gender, nationalism, and party politics. As I detail below, the precise form that these environmental dimensions of class struggle take has varied widely over time.

CLASS ECOLOGICAL ORDERS IN US POLITICS

Over the course of this book, I situate my analysis of US anti-environmental politics within distinct "class ecologies" that have evolved from the mid-nineteenth century onward.[3] I define a class ecology as the sum of the relations through which classes struggle to gain control over or access to particular forms of nature (e.g., land, bodies of water, minerals, food and energy sources), often relying on environmental concepts as they work to organize around shared identities, codify their gains in state policy, legitimate their victories, and rail against their losses. I refer to the institutionalization of a set of environment-class relations during a specific historical period as a "class-ecological order."[4] In the chapters to come, I provide sketches of three distinct class-ecological orders, which run from (1) the 1840s to the 1870s (what I term *classical class ecology*); (2) the 1940s to the 1970s (*middle-class ecology*); and (3) the 1980s to 2015 (*neoliberal class ecology*). These historical periods correspond with different regimes of accumulation that have structured global capitalism, with the United States playing very different roles in each. I conclude by suggesting that we are on the precipice of a fourth class-ecological order characterized by the reawakening of two historically consequential political forces, socialists and fascists, though this time with an environmental twist. Both eco-socialists and fossil fascists are seeking to hegemonize the relationship between class and the environment in an attempt to control the social and ecological processes so integral to their respective visions of the future.

In laying out these ideal-typical orders, my aim is not to present a comprehensive history of how class has intersected with the politics of nature (indeed, my analysis omits some of the most important eras of US labor struggles—the violent industrial clashes of the Gilded Age, the brutal crackdowns of the Red Scare, the advent of industrial democracy that occurred alongside the rise of the CIO, etc.). Rather, as an environmental political theorist, my intention is to develop a set of concepts that coalesce into a historically grounded theory of anti-environmentalism, one that enables us to think about the ways in which past struggles over the control of nature have shaped the present and to thereby more effectively intervene now and in the future. In the spirit of immanent critique, I attempt to take heed of the nightmarish weight of history while keeping the possibility open for dreams that could emerge out of the emancipatory kernels of our everyday lives.

The typologies that I present in the chapters that follow highlight deep-seated path dependencies that contemporary activists need to recognize in order to formulate effective strategies, as well as moments of rupture (both emancipatory and oppressive) that offer clues into points of systemic instability and structural vulnerability. In focusing on the interface between class and ecology, I am not suggesting that race, gender, nationalism, political party affiliation, religion, or generation play bit roles in political outcomes (to the contrary, my analysis suggests that they are integral); but I do place emphasis on capitalism as the most powerful driver of social and ecological outcomes in our current era, and I believe that working-class solidarity has the unique potential to overthrow an existing order beset by inequality and environmental crisis and to awaken the possibility of collective liberation for exploited humans and nonhumans alike. The reason for this is simple: the working class, properly conceived, encompasses the vast majority of people in the world, and—as a form of collective agency—it alone occupies the capacity to bring the global political economy, and the fossil fuel infrastructure that sustains it, to a sudden halt (Zweig 2012; Huber 2022).

My analysis is indebted to and builds on a burgeoning field of eco-socialist research that explores the history of working-class environmentalism (Gordon 2004; Barca 2012; Loomis 2016; Montrie 2008, 2011; Huber 2019) and labor-environmentalist relations (Dewey 1998; Obach 2004; Stevis 2011; Rector 2017, 2018; Räthzel, Stevis, and Uzzell 2021; Stevis 2023). However, perhaps because eco-socialists correctly assume that anti-environmentalism is part and parcel of the capitalist political economic order (see, e.g., Huber 2013; Malm 2016), there is comparatively little scholarship that systematically interrogates how class relations have factored into the logics, narratives, strategies, and policy prescriptions of organized anti-environmental movements and their political allies (notable exceptions include Bellamy Foster 1993; White 1996; and Malm and the Zetkin Collective 2021). Existing scholarship that focuses explicitly on anti-environmentalism is dominated by two approaches. The first is primarily empirically driven and descriptive in nature, analyzing the backlash to 1960s and early 1970s environmental laws, tracing a pathway from the Sagebrush Rebellion to the Heartland Institute (Jacques, Dunlap, and Freeman 2008; Dunlap and McCright 2010; Jacques 2012; Layzer 2012; Norgaard 2012; Brulle 2014; Farrell 2016; Turner and Isenberg 2018; Grasso 2019; Brulle 2022). The second is theoretically informed and normative in nature, reaching back centuries to locate an anti-environmental impulse that lies at the deepest roots of Western modernity—in Christian creationism, Cartesian dualism, or the Newtonian revolution (see, e.g., White 1967; Merchant 1980; Latour 1993; Moore 2015). The former focuses heavily on individuals and organizations who have actively fought against environmental laws, regulations, and movements; the latter tends to focus on the description and deconstruction of hegemonic discourses, or those that have attained a degree of dominance thanks to their ability to serve the interests of powerful actors. In the former, anti-environmentalism is the purview of political elites and grassroots activists; in the latter, it is woven into the material and ideological foundations of all of our lives.

Taking inspiration from both bodies of work[5]—but attempting to wed the tools of political theory to the former and a dose of historical contingency to the latter—I conceive of anti-environmentalism as a political project that opposes meaningful efforts to protect and preserve biophysical systems (including land, water, air, and the climate) and the health, habitats and homes of human and nonhuman species within these systems. Understanding the logics and strategies of the actors at the vanguard of putting anti-environmental ideas into action—for example, late -twentieth and early twenty-first-century right-wing movements, organizations, and parties—is, thus, a crucial facet of my research, but, as the Democratic Party of recent years has shown, policies that have profoundly negative environmental impacts increasingly march under green banners. Moreover, these contemporary actors work within historical paths paved by previous struggles over the control of land and the extraction of resources, and over the pollution of air, water, forests, and soil. I survey the evolving terrain of anti-environmentalism using multiple methods[6]—close readings of theoretical texts, archival research, participant observation, and interviews—and by engaging with an eclectic body of scholarship, including environmental history, labor history, critical race theory, American political development, ecofeminism, Marxist theory, and environmental justice studies. Over the course of the book, I trace the trajectory of anti-environmentalism from the nineteenth-century frontier to the 1950s suburb, from the shuttered shops of Main Street to the extractive economies of Trump country, with an eye toward understanding how ostensibly environmental signifiers have been deployed in the service of social hierarchy; how actually existing nonhuman lives, forces, and flows have been appropriated and channeled into circuits of capital and elite-driven schemes of statecraft; and how working-class movements have turned to nature as a source of material sustenance, spiritual inspiration, and political demands.

Chapter 1 examines an era during which the foundations for both the US working class and US conservationism were being laid: the mid-nineteenth century. To gain insight into the class-ecological order of the period, I begin by turning to perhaps the most astute scholar of mid-nineteenth century US politics: W. E. B. Du Bois. Du Bois was most influential for his penetrating insights into race-class relations, and his magnum opus, *Black Reconstruction in America* (1992), has spawned a whole genre of scholarship—whiteness studies—that is increasingly invoked to explain the rise of reactionary social forces today. I argue that one vital element of Du Bois's famous depiction of the "wages of whiteness" is almost entirely omitted from extant analyses: the struggle to control land and gain access to the fruits of nature. By putting Du Bois into conversation with social and environmental histories of the mid-nineteenth century, as well as primary research into abolitionist and labor periodicals of the era, I advance an alternative reading of *Black Reconstruction* and introduce the idea of the *natural wages of whiteness*. The economic and psychological wages of which Du Bois spoke also had an ecological counterpart that influenced the process of class formation.

Chapter 2 moves from the mid-nineteenth-century frontier to the mid-twentieth-century suburb. Popularly depicted as the locale where an emergent white middle class traded class struggle for consumerism in making its home at the end of ideology, the suburbs were, in actuality, intensely political spaces, where environmentalism (of a particular sort) flourished, but the ideological foundations were also being laid for anti-environmentalism. In order to understand the class ecological order of the mid-twentieth century, I weave together insights from two bodies of scholarship: *the treadmill of production*, which explains how nature and labor were respectively being decimated by the energy- and chemical-intensive realities of Fordist production (Schnaiberg 1980; Gould, Schnaiberg, and Weinberg 1996; Schnaiberg, Pellow, and Weinberg 2002; Gould, Pellow, and Schnaiberg 2004, 2008); and the aforementioned concept of *social reproduction*, which examines the relationship between labor and the gendered realities of mid-twentieth-century life (Dalla Costa and James 1972; Federici 1974). I argue that, in a regime of production predicated on a "family wage" for male breadwinners, the predominantly white, middle-class women of suburbia found themselves responsible for keeping up with the demands of an accelerating *treadmill of social reproduction*, in which the care work required to sustain a middle-class family spilled out beyond the home and into efforts to preserve neighborhood and community stability: for example, membership on school boards, parent-teacher associations, homeowners associations, and local anti-sprawl and conservation groups (Sugrue 1996; McGirr 2001; Geismer 2015).

As the post–World War II political economic order—and the industrial working class to which it was so resolutely tethered—reached its apogee in the late 1970s, suburban environmentalism found a national voice in the New Democrats (and the Beltway environmental groups with whom they allied themselves), while the logics, strategies, and institutional forms of the

New Right, now firmly ensconced in the Republican Party, were drawn into an anti-environmental movement with both elite and grassroots offshoots. At this moment—amid the ruptures of globalization and deindustrialization, the deterioration of the family wage, the ascent of the carceral state, and the institutionalization of a neoliberal political economic rationale—the rapidly shifting ground of jobs and employment left many across the political spectrum wondering what was to be of the working class. Lost in the noise of the Cold War and the culture wars, few noticed that the environment was gradually being woven into an evolving class politics of the American Right.

Chapter 3 surveys the class-ecological order of the era, arguing that, from the late 1970s onward, an emergent narrative, characterized by appeals to the heartland, middle America, Main Street USA, and the "Real America," was working to advance a lexicon of class that depicted workers as under attack by (1) a *new class* of environmental technocrats and (2) an *underclass* of the poor, marginalized, and unemployed. At the very moment that the Right's longstanding anti-union agenda was decimating organized labor, a new conservative iconography—proliferating images of coal miners, roughnecks, loggers, farmers, and ranchers—was being deployed to both appeal to workers and reinforce capital's attempts to evade environmental laws and regulations. As the power of finance capital surged, outsourcing and automation accelerated, and corporate profits skyrocketed, the anti-environmental movement elevated environmentally destructive labor into a national birthright and pitched environmentalism as a grave threat to US jobs. Since the turn of the century, this anti-environmental animus has only intensified; today, the Republican Party has turned to extractive laborers as the vanguard of the working-class activism that will "Make America Great Again" while portraying environmentalists as emblematic of professional class elitism. Weaving socio-ecological tropes from previous eras together into a defense of anthropo- and androcentric white nationalism, the Right attempts to cement the natural wages of whiteness into perpetuity through visions of high-paying extractive jobs for white male workers and continued environmental privilege among the suburban set whose anger and animus is integral to the surging tides of fossil-fueled fascism.

However, the efficacy of this strategy is often overstated in now ubiquitous depictions of a "white working class" wedded to Trumpism. In chapter 4, I review the literature on the (anti)environmental politics of so-called Trump country and then stage a debate between this body of work and environmental justice scholarship and activism. Drawing on interviews with environmental activists in "Trump country," I argue that while neoliberalism does indeed threaten to "undo the *demos*," as Wendy Brown (2015) has convincingly maintained, its concomitant undoing of the *oikos*, the socio-ecological place that we call home, is producing reactions that are transforming politics in parts of "Trump country." As the treadmill of production has become increasingly tethered to a global assemblage of financial and fossil capital, the treadmill of social reproduction has sped up to such an extent that it is reaching a breaking point: mounting housing and childcare costs; the plague of addiction; toxic contamination of land, water, and air; and the destabilization of everyday life by climactic crises. The resulting anger and outrage at the deterioration of home life has, in some cases, fueled insular modes of community activism seeking to secure the *oikos* from racialized Others while doubling down on extractive jobs as the path toward community revitalization (Hochschild 2016; Brown 2019). But, in other cases, the undoing of the *oikos* has provoked resistance against extractive projects and the formation of surprising new coalitions. At a moment when whiteness has been transformed into something of a master signifier around which many orient their analyses of reactionary politics, there exist deep cracks and fissures among actually existing white workers. The undoing of the *oikos* is steadily forging a shared set of interests between working-class communities, across lines of race, gender, and nationality, who lack the resources to distance themselves from contemporary environmental threats. This has already led to promising examples of collective action and the stirrings of solidarity. The environment, long integral to the wages of whiteness, is emerging as a prominent terrain of struggle on which they can be challenged. Only by doing so can the anti-environmental rage of the present be effectively surmounted.

CONCLUSION: PUTTING ANTI-ENVIRONMENTALISM TO WORK

The strength of anti-environmentalism is the extent to which it has been structured into our day-to-day lives. The outcomes of previous anti-environmental victories quite literally saturate our beings: from the jobs to which we have access to the ways that we produce and from the foods that we consume to the myriad synthetic chemicals permeating our bodies. This, however, is also anti-environmentalism's greatest weakness—the toxic by-products of this saturation are opposed by virtually everyone. Most people want jobs where they don't have to inhale dangerous amounts of contaminants, and they want to live in places where they can drink clean water, breathe decent air, and enjoy forests, fields, rivers, and lakes with their families and friends. The jobs versus environment trade-off has become common sense, in part, because the trade-off continues to exist for workers in some industries, with some states and regions particularly hard-hit. But with the right political interventions, this devil's bargain needn't persist.

Putting anti-environmentalism to work means embedding our analysis of environmental struggles within class politics, but it also means considering how the strategies, logics, and material interests of anti-environmentalism can be turned against themselves—by revealing the entire apparatus of organized opposition to meaningful environmental measures to be nothing more than a tool of capital: a strategy, born of desperation, that seeks to stall or stop the speed of climate mitigation and thereby control costs of production while dividing working-class movements at the very cost of our democracy and planetary health. The core vulnerability of fossil capital is that the number of fossil capitalists—those who actually benefit from continued extraction—is incredibly small; they require political allies outside of their ranks. In order for fossil capital to continue securing its spoils, fossil capitalists and their allies need to convince a sizable segment of the population that living and working amid the smoke is worth it. The reality is that the success of contemporary anti-environmentalism is predicated on its acceptance by workers—not only the well-worn tropes of conservative iconography but the janitors, maids, line cooks, secretaries, nursing assistants, teaching aids, adjunct professors, and many more people who are struggling to get by amid the inequalities and instabilities of the present.

It is no exaggeration to say that the future of the environment is in the hands of the working class. Perhaps the future of the working class also lies with the environment.

1 ROLLIN' COAL WITH DU BOIS: THE "WAGES OF WHITENESS" ON THE WILD FRONTIER

A few years ago, I found myself standing outside of a hotel in Flagstaff, Arizona, where a group of immigrants' rights activists had gathered to protest the establishment's mistreatment of undocumented workers. As we demonstrated alongside the main entrance, holding our banners and signs, a large pickup truck drove by, and we found ourselves immersed in a cloud of jet-black smoke. The truck sped off as the smoke lifted, along with several middle fingers from the crowd. At the time, I wrote this off as an isolated incident, but I soon learned that it is something of a phenomenon: *rollin' coal*, as it is known by its practitioners, has, among a small but apparently dedicated subset of mostly young white men, become a common tactic for harassing bicyclists, Prius drivers, and suspected "social justice warriors." The hobby—which involves modifying a diesel engine and installing custom exhaust pipes in order to emit visible plumes of smoke—reflects a subculture in which youthful acts of rebellion and enjoyment of automobility are quite literally filtered through the prism of pollution (while the sulfur dioxide being spewed into the air is not filtered through a catalytic converter). For the truck driver and his passenger, blowing exhaust on us was a means to express their anti-environmental and anti-social-justice attitudes in a single act.

This incident reflects a broader phenomenon in US politics: white identity politics, hypermasculinity, and a powerful image of working-class culture have become tightly tethered to anti-environmentalism. The national echoes of these localized outbursts became abundantly clear during the Trump presidency, where the connection between extractive jobs and the well-being of the working-class was conspicuously deployed to justify myriad anti-environmental policies: from executive orders that aimed to repeal the Stream Protection Rule and the Clean Power Plan to attempts to expedite the approval of new oil and gas pipelines to cuts in funding to agencies tasked with environmental protection. As Vice President Pence, surrounded by twelve white male coal workers, reiterated at a 2017 press conference unveiling the "Energy Independence" executive order: "President Trump digs coal" (ABC News 2017). At a moment where the Republican Party seeks to justify its inaction on climate change by appealing to the livelihood of extractive workers and equating any climate policy, however reformist, with "socialism"—the symbolic and material linkages between anti-environmentalism and the (white male) working-class remain prominent.

This chapter begins to unpack the history behind these linkages, and to contribute to a historically grounded theory of US anti-environmentalism, by turning to W. E. B. Du Bois's *Black Reconstruction in America* (1992 [1935]). Putting Du Bois, as well as more recent work in whiteness studies, into conversation with social and environmental histories of the Civil War–era frontier, I make the case that the "wages of whiteness" (which I explain in detail below) have always been constituted through the politics of nature. Not only have shifting ideas of nature been used to legitimate social hierarchies and thereby disable cross-racial working-class projects, but concerns that we would today understand as environmental—struggles over access to and control over land and natural resources, and anger over exposure to polluted air and water—were central to class and racial formation during a historical period that remains all too relevant to contemporary US politics.

While historical in its focus, this chapter is in dialogue with several contemporary areas of study that are of vital interest to environmental scholars and activists: a nascent body of critical environmental justice studies that has begun to explore the relationship between racial formation and environmental politics (Park and Pellow 2004; Washington 2005; Pulido 2015; Taylor 2016; Zimring 2017; Heynen 2018); recent scholarship that excavates instances of working-class environmentalism from a historical narrative that has tended to emphasize the movement's elite and middle-class histories (see, e.g., Gottlieb 2005; Montrie 2008; Rector 2014); a growing literature on anti-environmental politics and climate denial (see, e.g., McCright and Dunlap 2011, 2013; Brulle 2014; Malm and the Zetkin Collective 2021); and several works that extend Du Bois's insights on the nature of US racism into environmental politics (Smith 2007; Jobson 2021; Bhardwaj 2023). My argument is that examining the origins of what I term the natural wages of whiteness can give us insight into the historical grooves in which attempts to combat environmental degradation and achieve socio-ecological justice remain mired. However, in contrast with works that treat contemporary political identities as so sedimented as to be fossilized, my analysis also calls attention to the contingent historical processes through which whiteness has been resisted and reconfigured, in part, through the realities of environmental struggle.

THE WAGES OF WHITENESS

Du Bois begins *Black Reconstruction* by conceptualizing the struggle against slavery as a struggle between two labor movements—one led by abolitionists seeking to rid the nation of the evils of slavery, and another led by white workers seeking better wages and working conditions. The movements, however, failed to recognize their inner solidarity of interests (1992, 20–22). The rest of *Black Reconstruction* provides a detailed analysis of this failure. In an oft-cited passage, Du Bois referenced the wages of whiteness:

> It must be remembered that the white group of laborers, while they received a low wage, were compensated in part by a sort of public and psychological wage. They were given public deference and titles of courtesy because they were white. They were admitted freely with all classes of white people to public functions, public parks and the best schools. The police were drawn from their ranks, and the courts, dependent upon their votes, treated them with such leniency as to encourage lawlessness. Their vote selected public officials, and while this had small effect on the economic situation, it had great effect upon their personal treatment and the deference shown to them. (1992, 700–701)

The wages of whiteness, here, include material and ideological considerations that cut across political, economic, and cultural domains; entrance into the sphere of whiteness enabled access to better jobs and schools, granted rights of citizenship (like voting and some degree of participation in public debates), provided a privileged social status, and ultimately influenced the worldview adopted by white workers. Du Bois turns to the wages of whiteness only in passing—one paragraph tucked away near the end of a seven-hundred-plus-page book—but the early chapters of *Black Reconstruction* provide deep descriptions of the same phenomenon, and an entire genre of critical race scholarship, whiteness studies, has since employed and expanded on Du Bois's analysis to explore the fractured relationship between white workers and working-class people of color (seminal texts include Saxton 1990; Roediger 1991; Harris 1993; Ignatiev 1995; Foley 1997; Brodkin 1998; Lipsitz 1998; Hale 1999; Mills 2004; Olson 2004; more recent works include Painter 2010; Hosang and Lowndes 2019; Beltrán 2020).

The roots of whiteness studies can be traced back to internecine New Left squabbles of the 1960s, wherein white radicals in Students for a Democratic Society read Baldwin and Du Bois while trying to come to grips with the racism that persisted among white factory workers, and the role that white activists should play in struggles for civil rights amid the disintegration of SNCC and the rise of Black Power (see, e.g., Ignatin and Allen 1969). However, whiteness came to constitute an academic field of study only in the 1990s and early 2000s (Haider 2018; Johnson 2019). At the center of the field lies an attempt to build on Du Bois in order to explain the historical failures of the working class to coalesce around shared political projects and identities, and to interrogate the possibility for working-class solidarity today. Roediger, perhaps the best-known exponent of this field, explains that divisions in the working class can be tracked back to the construction and gradual reification of whiteness: "Working-class formation and the systematic development of a sense of whiteness went hand in hand for the US white working-class" (1991, 8).

In the decades since whiteness studies first came into being, however, the conceptual utility of the concept has been called into question by a number of leftists with long histories of theorizing and organizing against racism and white supremacy (Arnesen 2001; Fields 2001; Reed 2001, 2013; Johnson 2019, 2023). Analyses centering around whiteness, the critics argue, reify ascriptive racial categories as essentialized forms of difference; employ transhistorical understandings of whiteness that draw too smooth an equivalence between previous historical eras (e.g., slavery, Jim Crow) and the present; rely on a mistaken assumption that the "color line" is the central social antagonism for understanding inequalities across time and space; shift analytical focus away from how capital's drive to extract surplus value lies at the foundation of the multidimensional fractures within the working class; and couch their emancipatory hopes in allegedly "authentic" voices of color—from Du Bois to the Black Panthers—whose perspectives are taken to be synonymous with those of the whole race.

I find many of these criticisms persuasive, but the critics by and large fail to account for the internal variability within whiteness studies. To oversimplify for analytical purposes, there are two ideal-typical conceptions of whiteness. The first—running from Peggy McIntosh's account of "white privilege" (1989) through Robin DiAngelo's musings on "white fragility" (2018)—tends to use whiteness as a shorthand for the attitudes and behaviors of white people. This now-dominant approach, institutionalized in myriad DEI trainings and ubiquitous in popular media, recognizes that whiteness is a social construction but views it as so deeply entrenched that it has become woven into our individual and collective DNAs. Whiteness is a rarely visible but always present social fact, manifested in the dominance of white social norms, the prevalence of implicit bias, and the persistence of "microaggressions." Whiteness intervenes in and overcodes interpersonal relations, resulting in organizational decisions and institutional policies that reflect white interests. From this perspective, the harms of whiteness can only be minimized through an acknowledgment of white privilege that leads to introspection (at an individual level) and epistemic deference to non-whites (at interpersonal and organizational levels).

The second conception of whiteness flows from Du Bois, through 1960s radicals-cum-academics like Theodore Allen and Noel Ignatiev, and more contemporary scholar-activists like Asad Haider (2018) and Jeff Goodwin (2022). This approach views whiteness as an ideology that activates a racial identity among white workers, thereby creating the material and psychological incentives for a coalition between working class whites and capital. This approach does not deny the persistence of individual racism or the material realities that the ideology of whiteness has produced (such as continued racial disparities in health, wealth, and education), but insists that the forces driving social inequality are best understood through an analysis that places the evolving terrain of capitalism and class struggle at its center. Ultimately, adherents of this conception maintain that the ideology of whiteness can be overcome only through the construction of bonds of cross-racial solidarity forged in the fire of social movements.

While too many scholars flow back and forth between the aforementioned approaches without confronting the tensions between them (see, e.g., Coates 2017), they emerge out of fundamentally different ontological assumptions and political projects: the former reduces whiteness to a static identity and/or a transhistorical reality, the latter engages in historical analysis of the shifting grounds through which the cross-class coalition of whiteness is constituted; the former is firmly grounded in a liberal approach to anti-racism that focuses on individual attitudes and interpersonal interactions, the latter attempts to undertake a historical materialist "investigation into the ideology of race" (Haider 2018, 51); the former pursues organizational change through reflection and moral awakening, the latter attempts to identify the conditions of possibility for achieving working-class power and, with it, political economic transformation.

Beginning in this chapter, and continuing over the course of the book, I argue that one of the foremost reasons behind the failure of working-class solidarity—and environmental progress—in the United States has been the persistence of a cross-class coalition between a sizable segment of working-class white Americans and capital. Moreover, I detail how racism continues to enable capital to take working-class experiences of exploitation and immiseration and channel them in the direction of culture-war politics. Taking heed of the aforementioned critiques of whiteness, while drawing on insights from the historical materialist strand of whiteness studies, I aim to proffer an eco-socialist account of whiteness and to consider its relationship to (anti)environmental politics.

I conceive of whiteness not as an ontological certainty, static identity, or transhistoric social fact, but as a "ruling class social control formation" (Allen 2001, 2021)—an ideological edifice that emerged out of the efforts of capital to legitimate the violence of a plantation economy that relied on slavery and Native genocide, that was then internalized by many white workers as they "responded to a fear of dependency on wage labor and to the necessities of capitalist work discipline" (Roediger 1991, 13). Ignatiev, for example, considers how "the Catholic Irish, an oppressed race in Ireland, became part of an oppressing race in America" (1995, 2). The short explanation for this abrupt shift is that they became white, throwing in their lot with the pro-slavery Democratic Party rather than with the Black workers alongside whom they often lived and worked. This embrace of whiteness delivered material opportunities—opening up access to jobs and markets, as well as the ability to participate in democratic political life (for men anyway) (Ignatiev 1995, 3). The *wage* of whiteness thus refers to the advantages, perceived or real, that people who happen to be white obtain by acting as white social and political subjects—that is, by entering into collective life carrying the assumption that their racial identity is central to their existence (Painter 2019). Of course, for the white workers, these advantages are only ever partial, contradictory, and self-defeating, ultimately serving to sever ties of class solidarity and thereby lock all workers into relations of exploitation.

Du Bois makes the case that, in an environment in which exploitation and suffering was the norm, white workers nonetheless had political choices available to them: in entering into a cross-class alliance with other whites rather than a cross-racial alliance with other workers, they chose the benefits of whiteness over those of working-class solidarity. Whiteness studies suggest that white Southern workers, mid-nineteenth-century German and Irish immigrants, early twentieth-century Southern and Eastern European immigrants, mid-twentieth-century white factory workers (and so on), found themselves in a world not of their own choosing, but where some political agency still existed. The puzzle laid out by whiteness studies—"Why would European-American workers respond to their exploitation and social degradation in the particular form of 'white' identification, rather than in 'non-racial' ways?" (Allen 2001)—has exploded into popular consciousness in recent years, as a tool for interrogating the election of Trump and the resurgence of the Far Right (see, e.g., Anderson 2017; Coates 2017; Metzl 2019). To what extent have concerns that we today recognize as "environmental" factored into the construction and maintenance of the white wage? And what use is the concept in understanding contemporary anti-environmental politics?

Whiteness and Anti-Environmentalism

Amid the persistence of environmental racism and the "spectacular racism" of the Trump administration (Pulido et al., 2019), environmental justice scholars have sought to further unpack the relationship between environmental injustice and racial formation—that is, "the sociohistorical process by which racial categories are created, inhabited, transformed, and destroyed" (Omi and Winant 1994, 55). A core argument of this body of work is that racial difference has been, and remains, central to the production of environmental inequality, and vice versa. For example, one key component of racial formation has been the construction of particular populations as biologically or culturally closer to nature and thus further from civilization. Such essentializations have been historically deployed to suggest that racial minorities are more inclined or adaptable to environmentally hazardous sites of residence and types of work (Park and Pellow 2004; Washington 2005; Finney 2014; Rector 2014; Pulido 2015; Zimring 2017), as well as to legitimate the theft of land and resources from indigenous nations and communities of color (Voyles 2015; Purifoy and Seamster 2021). These linkages between nature and race have intersected with other social hierarchies, such as gender and class, contributing to the reduction of care work to unpaid or underpaid domestic labor, the patriarchal domination of a feminized nature, and the intensified exploitation of workers of color.

While racial formation theory has been deployed to understand the ideologies that have perpetuated environmental injustices, considerably less theoretical attention has been paid to the relationship between anti-environmentalism and the construction of whiteness (exceptions include Washington 2005; Pulido 2015; Zimring 2017; Malm et al. 2021). In the discussion that follows, I argue that understanding and, ultimately, combatting contemporary anti-environmentalism requires an approach to whiteness that is attuned to both the path dependencies that have resulted from the outcomes of previous struggles, and the lessons that can be gleaned from historical moments in which unexpected solidarities emerged and emancipatory possibilities were revealed, however fleetingly. In *Black Reconstruction*, Du Bois offers a finely grained historical analysis that understands whiteness as a contingent political project situated within the evolving structure of capitalism and at the intersections of race, class, party, region, and nature. While his magnum opus does not fully incorporate the experiences of all workers—like women and Native Americans—into the book's conceptual apparatus (Hartman 2016; Estes 2019; Bruyneel 2021),[1] complementing Du Bois's insights with more recent scholarship on the era allows us to begin mapping out the historical trajectory through which the environmental dimension of the wages of whiteness—what I term the *natural wages of whiteness*—came to animate US anti-environmentalism. My analysis begins with the frontier—"the place," as historian Richard White put it, "where whites met the wilderness" (1996, 397).

THE WAGES OF WHITENESS ON THE WILD FRONTIER

In *Black Reconstruction*, Du Bois focused considerable attention on how the contentious politics of the frontier mediated social-ecological relations of the era. For the white planter class, the continuation of slavery required more and more land (1992, 41), and the possibility that this land could become "free" represented an existential threat to the "peculiar institution" on which its economic and political clout rested. Land in the west also represented an opportunity to improve living and working conditions for poor Southern whites, but they vociferously insisted that a comparable opportunity not be extended to freedmen (1992, 28). For some Northerners, the frontier represented a "safety valve" for class tensions that had begun to simmer after the Jacksonian-era extension of the franchise to non-property-owning white men. In a different respect, the frontier was also a safety valve for recent immigrants: a chance to escape the drudgery of the cities, reconnect with the land, and transform themselves into the yeoman farmers of Jeffersonian lore (1992, 23).

Nature played a crucial role in the construction of these social relations: it was deployed by slaveholders to legitimize racial hierarchy and by abolitionists to challenge that hierarchy; wilderness was a space to be tamed through labor, and a space of awe through which the nation could be revitalized; and the open and "unsettled" land of the west represented an opportunity for overcoming class conflicts and a siphon for channeling cheap nature and cheap labor into transnational circuits of capital.

The Right to Work the Wilderness

Philosopher John Locke famously turned to the wilderness of North America, arguing that nature not transformed by labor is wasted. The ability to put nature to use through labor was predicated on and legitimated the institution of private property. Native Americans also mixed their sweat with the soil (Cronon 1983; Blackhawk 2006; Dunbar-Ortiz 2014), but their property arrangements were scarcely intelligible to colonial settlers, much less political philosophers who had never visited the continent. The classical liberal theory of adding value to an otherwise worthless nature through labor was one that meshed with both reactionary and progressive movements of nineteenth-century America—wilderness not transformed was wasted, and more-than-human forces and lives were resources to propel human advancement. As an 1865 article in the abolitionist periodical *The Liberator* argued, "The capacity of the West for productiveness and population fills one with astonishment; and its future, in all that pertains to freedom and civilization, is full of glorious promise" (Anonymous 1865). The ability to put the land to work—the *right to work the wilderness*—was granted selectively to white males; European men were positioned as the sole community capable of transforming nature through labor, and therefore the only population fit for entrance into the national polity.

In spite of this inequality sewn into the very fabric of democracy in America, the supply of land and raw materials provided by the frontier was assumed by many commentators of the era to guard against the inequitable class relations that prevailed in Europe: "It soon became evident," noted Tocqueville, "that the soil of America rejected absolutely a landed aristocracy" (2003, 40). During the Jacksonian era in which Tocqueville roamed the country, a new democratic imaginary had emerged, replacing the ideal Jeffersonian citizen, the yeoman farmer, with the independent producer or "freeman." In the popular imaginary, a "producer," anyone whose wage labor was directed toward something deemed socially valuable, existed in contrast to a "parasite," who accumulated wealth through financial speculation and the work of others (Bridges 1986, 165). In this sense, the producer class was construed broadly, following an almost "catholic inclusivity" (Davis 1986, 14) that emphasized economic independence, and "drew no distinction between the laboring class and what we would call the middle class" (Foner 1995, 15). However, this class of *free men* existed in sharp contrast to both the *unfreedom* of chattel slavery and the unwaged domestic labor of *women*.

As industrialization intensified, the ideology of producerism was besieged by the realities of waged labor, in which workers did not possess economic independence but were bound up in intense and asymmetric relationships with the owners of the means of production (Roediger 1991, 45). This spurred the emergence of a Working Men's movement, as well as parties and labor unions in which a sense of class-for-itself began to develop—sometimes in conjunction with an interest in transforming what were seen to be inequitable and unhealthy nature/society relations. Du Bois finds that "in the United States shortly before the outbreak of the Civil War, there were twenty-six trades with national organizations, including iron and steel workers, machinists and blacksmiths" (1992, 25). The emergent (white) working class expressed concrete demands for "better public education, a ten-hour work day, abolition of contract labor on public works, lien laws, homestead legislation, and democratizing reforms (e.g. election of judges)" (Bridges 1986, 182).

Workplace struggles also frequently revolved around the "unnatural" environment of the industrial factory and the urban setting in which it was located. Historian Amy Bridges notes that "bakers claimed they needed organization to decrease the hours of work lest they be too weary after their 'week's unnatural toil' to study Nature and Nature's God" (Bridges 1986, 177). "Is a child born to liberty," an 1844 *Working Man's Advocate* article demanded, "who is subject, from infancy, to the loathsome contamination of a crowded city?" (Anonymous 1844). A Working Men's meeting in Otsego County (New York) stipulated that inalienable rights include "the right of every person to a sufficient quantity of light, air, land, water, and all the other bounties of nature, on and out of which to obtain, by his labor, a comfortable subsistence for himself and children" (Frink and Butler 1844).

This working-class ecology struggled against a powerful ruling-class ecology: intensified manufacturing required enormous transfers of natural resources from America's hinterland, in both the South and the West. Hornborg estimates that British manufacturers alone imported 223,623 tons of raw cotton from the US South in 1850 alone (which "liberated over 6 million hectares in Britain that could be turned to food production for the industrial labor force") (2006, 77–79; see also Johnson 2013, 10–11). Zimring writes that "demand by the mills of New England, the mills of South Carolina, and the mills of London led to vast expanses of land in Virginia, North Carolina, South Carolina, Georgia and moving west through Alabama and Mississippi becoming devoted to giant cotton plantations" (2017, 27). Industrial production was enabled by a system of coercive labor and by the ecological processes that had long unfolded in the Mississippi River Delta, giving rise to the fertile soils of the "black belt," where "the Mississippi had unloaded its rich sediments for millennia" (Beckert 2015, 113).

Of course, the land on which this agricultural production occurred wasn't empty. The expansion of the frontier required the forced relocation of its native inhabitants (see Beckert 2015, 107–108; see also Dunbar-Ortiz 2014; Taylor 2016; Estes 2019; Grandin 2019). Removal of Native American nations—including the Creeks, Cherokee, Chickasaw, Choctaw, and Seminoles—from the southeastern frontier opened the land to new forms of economic domination, justifying the clearing of wilderness for monocrops like sugar and cotton, which gave impetus to the relocation of approximately one million enslaved peoples to the Deep South (Taylor 2016, 125). The land was not enough, either; "the project of continental consolidation . . . also secured major rivers needed to carry the cotton," including the Mississippi, Red, Tombigbee, and Mobile (Beckert 2015, 108; see also Johnson 2013). As Du Bois described this historical moment:

> The giant forces of water and of steam were harnessed to do the world's work and the black workers of America bent at the bottom of a growing pyramid of commerce and industry; and they not only could not be spared, if this new economic organization was to expand, but rather they became the cause of new political demands and alignments, of new dreams of power and visions of empire. (1992, 5)

The Nature of Class (De)Formation

Under such circumstances, it would have made sense, per Du Bois, for working people of all races, genders, and nations to join forces against the ruling class to resist a system that worked for none of them. The US working class, however, was splintered—across lines of nationality, race, gender, and region—in ways that hindered the construction of strong bonds of solidarity (Davis 1986, 17–18). The ways in which the politics of nature factored into each of these antagonisms has been well documented, but are worth reexamining nonetheless.

First, the right to work the wilderness was simultaneously a right to dominate those populations deemed *part of wilderness* or viewed as *uncivilized by nature*. Native Americans were also excluded from the emerging conception of an independent producer, in large part because they so frequently resisted the institution of private property to which the concept was hitched (instead viewing land, water, trees and animals as kin with whom we are enmeshed in reciprocal relations) (see, e.g., Kimmerer 2013; Whyte 2016). The age of Jackson was the age of "Indian removal" and genocide. As John Ross, chief of the Cherokees, remarked in an 1836 letter to Congress, "We are denationalized; we are disenfranchised. We are deprived of membership in the human family!" (Beckert 2015, 108).

This ideology inflected even the "progressive" political movements of the time; an 1837 article in *The Liberator* referred to the "emigrating Indians" as "children of nature" and "sons of the forests" who were waiting out "the rising of the waters of the Arkansas . . . to enable them to reach their appointed homes in the far west." Such an attitude, even when intended to oppose the atrocities of Indian removal, placed Native Americans outside of civilization and the national polity. Proponents of removal, on the other hand, often justified the politics of conquest by appealing to the bountiful nature of the west. As Andrew Jackson wrote to the Cherokees in 1835, "The United States have assigned to you a fertile and extensive country, with a very fine climate adapted to your habits" (cited by Taylor 2016, 119). This rhetoric of natural abundance existed in sharp contrast to the reality—that Native Americans were typically pushed onto lands assumed to be barren wastelands, devoid of value (see, e.g., LaDuke 1999; Blackhawk 2006; Dunbar-Ortiz 2014; Voyles

2015; Black 2018; Estes 2019). As many indigenous scholars have pointed out, Native American communities continued to actively resist imperial rule and to creatively adapt to their new environs, but their subsistence practices and traditional socio-ecological relations were repeatedly undermined by further settler colonial expansion and the appropriative designs of belligerent extractive interests.

Second, the (white) working class was gaining a sense of class-for-itself at the very moment that Blacks were being stripped of their political agency. Du Bois points out that, "in the North, Negroes . . . received political enfranchisement with the white laboring classes" (1992, 7–8); however, this enfranchisement was gradually curtailed over the first half of the nineteenth century: "Northern Blacks were everywhere excluded from the universalization of manhood suffrage in the 1820s and 1830s, and on the eve of the Civil War only four states in the Union allowed freedmen even a qualified franchise" (1992, 27). Eric Foner makes the case that this relationship between the development of a working-class consciousness and the strengthening of slavery was not incidental; free labor ideology "could not develop without a sharpening of the actual dichotomy between slavery and freedom" (1995, xi). "All labor republicans," writes David Roediger, "existed in a society that offered the opportunity for white workers to measure their situations not only against the dream of a republic of small producers but also against the nightmare of chattel slavery" (1991, 44; see also Grandin 2019, 58).

The unfreedom experienced by putatively free Blacks extended into the economic realm as well: "Although they had been recently employed under slavery in a variety of skilled as well as unskilled occupations, emancipated Negroes found their economic opportunities limited to jobs as servants, seamen, or common laborers" (Allen 2001, 15; citing Litwack 1961, 154–155). Ignatiev adds that "black workers, already being driven out of artisanal trades by prejudice, and squeezed out of service trades and common labor by competition, could find no refuge in the manufacturing area, and hence were pushed down below the waged proletariat, in the ranks of the destitute self-employed: ragpickers, bootblacks, chimneysweeps, sawyers, fish and oyster mongers, washerwomen, and hucksters of various kinds" (1995, 134). While "working men" railed against the environmental ills of industrializing urban life as part of the drudgery of "wage slavery," such sympathy was rarely extended to Black workers, who frequently labored in "dirty" jobs, exposed to dangerous fumes, workplace safety hazards, and industrial toxins on a regular basis. A labor conference at Feneuil Hall in Boston in 1861 captured the general sentiment of early unions on slavery: "It is a matter which does not concern us" (Du Bois 1992, 25).

Enslaved peoples and their abolitionist allies—Du Bois's second labor movement—nonetheless drew on nature for material sustenance, spiritual guidance, and ideological legitimation (Glave and Stoll 2006; Smith 2007, Fiege 2012). Sentiments from the blossoming romantic movement often found their way into abolitionist attacks on slavery. Issues of anti-slavery periodicals, like *The Atlantic*, *The Liberator*, and *The North Star*, were littered with dirges to romantic nature, often juxtaposed with exposés detailing the unnatural and barbaric institution of slavery.[2] For example, a poem titled "To Nature," appearing in *The Liberator*, connected the pain of slavery with the pleasure of nature:

> There is too much of brightness in the sky, too green the earth all bathed in radiance seems.
> Too fair and flowery are the fields that lie, beneath the splendor of the day—god's beams.
> What though our country racked with inward pain, lifts up her voice with deepest anguish filled?
> What though the tears of the oppressed, like rain, Bedew the earth their own rough hands have tilled . . .

This relationship between slavery and wilderness was more than metaphoric. Dorceta Taylor points out that enslaved peoples "relied on nature and the wilderness for their sustenance and survival" (Taylor 2016, 127), and also saw it as a refuge from the violent oppression of the plantation. Harriet Tubman, for example, not only drew on her knowledge of nature (salves, birdcalls, how to move silently through the forest, how to feel the barks of trees for moss and tell direction, etc.), she prepared for her raids by lying alone in the forest all night (Taylor 2016, 136–139). Enslaved peoples sang songs about the wilderness that contrasted sharply with dominant portrayals of wilderness as foreboding (Dixon 1987, 12–13; cited by Taylor 2016, 139):

> If you want to find Jesus, go in de wilderness
> Go in de wilderness, go in de wilderness
> Mournin' brudder, go in de wilderness
> I wait upon the Lord.

It was this very song that "greeted the commander of the first slave regiment mobilized by the Northern side . . . [as they] marched through the country-side of South Carolina" (Malm 2018, 24; citing Higginson 1870, 133–135). Areas popularly imagined as wilderness—sparsely inhabited forests, mountains, and swamps—also contained the promise of relative freedom for run-away slaves, maroon communities, and free African Americans in the North (Blum 2002; Smith 2007; Diouf 2014; Golden 2021).

Third, the idea of a "*freeman*" also had an obvious gendered connotation. As Roediger notes, "In an urban society in which work and home became more radically separated and masculinity underwent extensive redefinition, its masculine ending may have had special appeal" (1991, 55). This exclusionary definition of productive labor is particularly ironic since the early working class largely comprised female laborers. Alice Kessler-Harris goes so far as to argue that "America's lack of an adequate supply of workers and ongoing need for cheap labor required that women become *the first industrial proletariat*" (2003, 22; emphasis added). Women did paid labor in their homes as well as in rapidly expanding mills and factories:

> By 1840, 37% of America's workers earned their livings outside agriculture. Less than 10 percent of these were in manufacturing. But women constituted half of the total and sometimes as many as 90% of those employed in shoe factories, textile mills, and millinery shops. (Kessler-Harris 2003, 29)

This first industrial proletariat also comprised enslaved women in the fields and plantation houses. Enslaved women were not only consigned to toil in plantation agriculture, but were also forced to reproduce to serve the economic (and, sometimes, sexual) ends of their owners. As Deborah Gray White notes,

"The slave woman's 'marital' status, her work load, her diet all became investment concerns of slaveholders, who could maximize their profits if their slave women had many children" (White 1985, 68; see also Berry 2017).

In spite of the centrality of women's work to the industrial economy, such a conjuncture was marked by a profound ideological shift in which middle-class white women were increasingly seen as guardians of the household—newly viewed as a refuge from the factory or from the financial stressors facing farmers who now depended on a steady profit to purchase (and pay debts on) manufactured goods and equipment. It is in this period that the "widely accepted distinction between productive and domestic work" emerged, a development that rendered invisible the work of women outside the home and devalued the work of women within it:

> The vaunted independence of the yeoman household depended in considerable measure on the labor of women—whether unremunerated within the household, or paid wages at home or outside. So too did the free time that enabled men to participate as citizens in the public arena. (Foner 1995, xxix–xxx)

Working-class women here faced a double bind: for working-class families, women's labor was a necessity for survival, but such labor was stigmatized by a dominant middle-class ideology that viewed time spent on labor as a threat to "good mothering and householding." For a woman, "working for wages was not simply demeaning, as for men, but *fundamentally alien to her nature*" (Foner 1995, xxix; emphasis added).

Fourth, the gender norms of the mid-nineteenth century meant that middle-class women were increasingly confined to the home, but the factories found a new source of workers in European immigrants. As Du Bois noted, "The growing exploitation of white labor in Europe, the rise of the factory system, the increased monopoly of land, and the problem of the distribution of political power, began to send wave after wave of immigrants to America, looking for new freedom, new opportunity and new democracy" (Du Bois 1992, 5). Producers were split between mostly native-born artisanal workers and predominantly immigrant wage workers. Unsurprisingly, "the foreign born were concentrated in the most industrialized, least skilled, and least secure employments" (Bridges 1986, 173). Among other perilous tasks, this new "'army of wage workers' dug America's canals, built the railroads, and loaded and unloaded ships in port cities" (Foner 1995, xv).

In contrast to African Americans, women, and Native Americans—who would have to wait to be welcomed into the ranks of the working class—European male immigrants found a side door into the exclusionary "house of labor." The immigrants came overwhelmingly to cities at the moment when midwestern statehood was being achieved. This, combined with Southern over-apportionment due to the three-fifths compromise, was a tremendous disadvantage to the electoral power of Northern urban residents. Since "wage labor was largely confined to the cities and nationally to minority status, . . . the same frontier that welcomed the pioneer and the farmer made of the working-classes an urban minority" (Bridges 1986, 161, 168). This meant that even though the demands of the working classes meshed uneasily with the political platforms of either major party, for urban industrial workers to exercise political power at the state and federal levels, they needed to work through one of the existing party apparatuses.

The parties responded to this gradual transition to wage labor in distinct ways: for the Democrats, the solution to the rise of industrial labor lay in recruiting whites into a *herrenvolk* democracy—citizenship protections for white workers, including immigrants, but slavery and conquest for nonwhites; for Republicans, the answer lay in free labor and the free soil of the frontier—tethered, for many, to a nativism aiming to exclude recent immigrants from political participation. The industrial behemoth could be sustained by free land, while the army of wage labors that it was rapidly producing could be transformed into free men by the opportunities of the frontier. As George Henry Evans put it, "We simply propose that . . . the land shall be left, as Nature dictates, free to the use of those who choose to bestow their labor upon it" (1844). The basic Republican answer to the problem of urban poverty, wrote Foner, "was neither charity, public works, nor strikes, but westward migration of the poor aided by a homestead act" (1995, 27).

The case of the Irish is instructive here. For both Du Bois and subsequent whiteness scholarship, the Irish factor prominently into the history of racial antagonisms among the working class. Irish Immigrants in the North were generally in a dire situation, facing low wages and harsh working conditions.

They also faced sustained nativist attacks and were at times viewed as a separate race entirely: "Strong tendencies existed in antebellum America to consign the Irish, if not to the black race, then to an intermediate race located socially between black and white" (Ignatiev 1995, 89). In spite of pleas from abolitionist leaders, like Frederick Douglass, Daniel O'Connell and Wendell Phillips, to throw in their lot with enslaved peoples and free African Americans and fight in solidarity against the exploitation of their labor, the Irish found another way out of their social position. As David Roediger notes,

> The Catholic Irish, the immigrant group most exposed to nativist opposition, accepted protection from Democrats. . . . Democrats and Irish-American Catholics entered into a lasting marriage that gave birth to new ideologies stressing the importance of whiteness. (1991, 141)

"By the mid-1840s," according to Nell Painter, "Irish American organizations actively opposed abolition with their votes and their fists" (2010, 143). *White standing* gave the Irish the right "to enter labor market as free laborers, . . . compete for all jobs, . . . vote and be elected, . . . hold positions within Democratic party machine, be tried by a jury, . . . live wherever they could afford," and so on (Olson 2004, 46). It also gave them the right to pursue opportunities provided by the land and resources of the west.

And yet, in contrast to the white middle class, this right to move west did not translate into much actual migration—perhaps, in part, because of a lack of interest, but more likely because of a lack of financial resources:

> Unable or unwilling to avail themselves of the white-skin privilege of setting themselves up as independent farmers, the vast majority clung to the Democratic Party, which continued to protect them from the nativists and guarantee them a favored position over those whom they regarded as the principal threat to their position, the free black people of the North. (Ignatiev 1995, 102)

Du Bois concludes, "The union and labor leaders gravitated toward the political party which opposed tariff bounties and welcomed immigrants, quite forgetting that this same Democratic Party had as its backbone the planter oligarchy of the South with its slave labor" (Du Bois 1992, 19). "Native Protestant workers," according to Mike Davis, "rallied to the leadership of their

Protestant bosses and exploiters, while Catholic immigrants forged an unholy alliance with Southern reaction" (1986, 27).

This Southern reaction was sculpted with the aid of white workers. In contrast to the intense and widespread efforts to integrate the Irish into party politics, white workers of the South, Du Bois pointed out, were largely ignored—by the labor movement, by the abolitionists, by the Northern capitalists and by Southern planters (1992, 26). The three-fifths compromise not only reduced slaves to less than personhood and put new immigrants in a quandary, it also disenfranchised poor whites, insofar as "the slaveholder practically voted both for himself and his slaves." In an 1861 article in *The Atlantic*, Charles Francis Adams Jr. highlighted the disconnect between the abundant natural resources of the South and the continued poverty and underdevelopment of the region:

> That the whole policy of the Cotton dynasty tends necessarily to making broader the chasm between these orders is most apparent. It makes the rich richer, and the poor poorer; for, as, according to the creed of the dynasty, capital should own labor, and the labor thus owned can alone successfully produce cotton, he who has must be continually increasing his store, while he who has not can neither raise the one staple recognized by the Cotton dynasty, nor turn his labor, his only property, to other branches of industry; for such have, in the universal abandonment of the community to cotton, been allowed to languish and die.

This relationship between slavery and the degradation of *both land and people* was a common observation. William Seward found in an 1846 trip to Virginia that "the land was sterile, the fences mean, and a universal impress of poverty stamped on all around me" (Foner 1995, 41; see also Olmsted 1861; Bagley 1942). Du Bois similarly juxtaposed "the endless land of richest fertility, natural resources such as Earth seldom exhibited before" (1992, 29) with the baffling inefficiencies and inhumanity of the Southern plantation owner who "wanted large income without corresponding investment and . . . insisted furiously upon a system of production which excluded intelligent labor, machinery, and modern methods" (1992, 37).

This "forgotten mass of men," as Du Bois refers to white workers of the south, were nonetheless political agents in their own right, and were not monolithic. On one hand, a "middle class of poor whites in the making" (overseers, merchants, police, etc.) gained materially from slavery. On the other, a "revolt of the poor whites" emerged in the form of western migration to the frontier. Southern whites, in both camps, were vehemently opposed to abolitionism, particularly in the western territories. The promise of the frontier not only failed to serve as a safety valve for racial antagonisms, it amplified them:

> Here was a solution such was impossible in Europe: plenty of land, rich land, land coming daily nearer its own markets, to which the worker could retreat and restore the industrial balance ruined in Europe by the expropriation of the worker from the soil. . . . This thought, curiously enough, instead of increasing the sympathy for the slave turned it directly into rivalry and enmity. (Du Bois 1992, 19)

Du Bois notes that, "while the Northeast demanded free soil, the Southerners demanded not only free soil but the exclusion of Negroes from work and the franchise" (1992, 28). According to Richard White, Southerners feared that such policies would "increase the number of 'free farms with Yankees and foreigners pre-committed to resist the participancy of slaveholders in the public domain' . . . 'Better for us, declared a Mississippian, 'that these territories should remain a waste, *a howling wilderness*, trod only be red hunters than be so settled'" (White 1991, 140; emphasis added). As one resident of Kansas put it, "We will continue to lynch and hang, to tar and feather and drown every white-levered Abolitionist who dares to pollute our soil" (Du Bois 1992, 42).

Abolition-Democracy versus Ruling-Class Ecology

The Civil War produced a sea change on the frontier, but one that failed to live up to the emancipatory hopes of abolitionists and Free Soilers alike. Without the Democratic opposition in the South, the Northern government was finally able to push through the Homestead Act of 1862, which granted 160 acres to western settlers who agreed to "improve" the land. Historian Keri Leigh Merritt calls this "the most comprehensive form of wealth redistribution that has ever taken place in America":

> The law granted a gigantic land mass—close to the size of both California and Texas—to more than 1.6 million households. These settlers, of courses, were almost exclusively white. (2017, 38)

Dreams of the frontier loosing forth a mass of small-scale independent producers, however, didn't come to pass. As Greg Grandin writes, "Within a decade of the act's passage, large capitalists and speculators had laid claim to the most fertile, best irrigated, and via railroad lines, best connected portion of public 'free land'" (2019, 110). Native American nations lost nearly 300 million acres of land (Dunbar-Ortiz 2014, 141), while the railroads were prime beneficiaries: Union Pacific and Central Pacific received twenty odd-numbered sections of land for each mile of track they constructed, and Northern Pacific acquired about 40 million acres (White 1991, 145–146). Cattle barons also made out like bandits: "They falsified titles using the signatures of cowhands and family members, employed fictitious identities to stake claims, and faked improvements on the land to appear to comply with the law" (Ketcham 2015, 25). The intensity of such duplicity on the part of bankers, developers, and mining, timber, and railroad interests led one historian to conclude that "probably most private range land in the western states . . . was originally obtained by various degrees of fraud" (Ketcham 2015, 25). Du Bois wrote that "the West, instead of becoming a country of peasant proprietors who might have counteracted this result, surrendered itself hand and foot to capitalism and speculation in land" (1992, 48).

For those small-scale producers and laborers who did settle on the frontier, the new environment often proved daunting, and many blamed the harshness of nature for their difficulties: "They had trusted nature and when nature behaved according to its own rules and not theirs, they felt betrayed" (Limerick 1988, 42). Patricia Limerick notes that nature, "Indians," and the federal government were common scapegoats in this setting (1988, 43–45), as were other populations who found themselves at odds with an increasingly dichotomous racial order. For example, the Hispano communities of the Southwest existed against the historical backdrop of a double colonization—Spanish and American—and had to contend with the subordination that "Mexicans" faced within a newly dominant Anglo racial imaginary (Gómez 2007). While their rights were supposed to be guaranteed by the Treaty of Guadalupe Hidalgo, efforts to square communal rights to land and resources with an American property regime resulted in theft and dispossession: "Between 1854 and 1891, only twenty-two of the more than two hundred Spanish land-grant claims were verified by the court, leaving 35 million acres of New Mexico's land unadjudicated" (Kosek 2006, 40). Where these claims did enter into judicial proceedings, supposed ambiguity over individual property rights was routinely cited as a basis for rejecting Hispano land rights, to the benefit of white elites. In short, the wild frontier failed to provide a safety valve for class or racial strife; rather, it served as a staging ground for the aggravation of racial divisions within the working class, as well as the appropriation of the common resources that paved the way for capital's reign during the Gilded Age. The western farmers who might have joined in common struggle with Black workers wanted the "freedom to exploit as well as to vote" (Du Bois 1992, 216).

In spite of a similar ending, the South presented a different story. The enslaved peoples' central position within the production of food and raw materials meant that they occupied a choke point in the global political economic order; their refusal to work the land could impede the flow of commodities on which both capitalist profit and the war effort depended (Du Bois 1992, 63–64, 121). Du Bois conceptualized the refusal of enslaved peoples to work and their eventual migration to join the Union Army as a *general strike*, but one that left the site of production in search of freedom and, eventually, further political engagement. The formerly enslaved were striking workers, in this sense, but they were also refugees ambling toward collective action:

> As a final gesture, they marched with Sherman from Atlanta to the sea, and met the refugees and abandoned human property on the Sea Islands and the Carolina Coast. . . . This was not merely the desire to stop work. It was a strike on a wide basis against the conditions of work. It was a general strike that involved directly in the end perhaps a half million people. They wanted to stop the economy of the plantation system and to do that they left the plantations. (Du Bois 1992, 67)

It was here that General Sherman's Field Order Number 15 (also referred to as the Sea Island Circular) provided a model of reclaiming the commons, a very small repayment in nature for loyalty to the union cause:

> The islands from Charleston south, the abandoned rice fields along the rivers for thirty miles back from the sea and the country bordering the St. John's River, Florida, were reserved for the settlement of the Negroes made free by the acts of war and the proclamation of the President. (Du Bois 1992, 73)

Altogether, 485,000 acres were divided among 40,000 formerly enslaved peoples who had followed the Union Army (Du Bois 1992, 393). In subsequent years, according to General Rufus Saxton, the bountiful production on the lands of former slaveholders fed the communities and produced additional goods for market (Du Bois 1992, 74).

The expropriation of land owned by slaveholders and the redistribution of that land to the newly freed was a high-water mark in the broader project of abolition-democracy. The immediate demands of Black workers, according to Du Bois, were land and education, the provision of sustenance for the body and the mind (1992, 123), which they sought to guarantee through a concomitant struggle for the right to vote. With the plantation infrastructure torn asunder, and the plantation class attempting to resurrect the antebellum status quo, Northern industry made a tactical alliance with abolitionists, labor, and the cause of freed slaves, establishing a temporary *dictatorship of labor* that pressed for education, land reform, and civil rights for Blacks (Du Bois 1992, 184–185).

A vital cog in the gears of abolition-democracy was the establishment of the Bureau of Refugees, Freedmen and Abandoned Lands (the Freedmen's Bureau, for short), which was tasked with providing food to the hungry, land to the homeless, opening schools and hospitals, and protecting the political and economic rights of Blacks living in the midst of an angry white citizenry (Du Bois 1992, 225–227). A target of hostility and outrage from the start, the bureau was plagued by chronic underfunding and understaffing; nonetheless, it was, for Du Bois, "the most extraordinary and far-reaching institution of social uplift that America has ever attempted" (1992, 219). The importance of land—of nature—to the promise of reconstruction cannot be overstated; the freed peoples' "demand for a reasonable part of the land on which they had worked for a quarter of a millennium was absolutely justified, and to give them anything less than this was an economic farce" (Du Bois 1992, 602). The cry of "land for the landless"—a cry that was frequently invoked by Black leaders to apply to the landless of all races—revealed the centrality of nature to the material and spiritual sustenance of human life, and the ways in which the working class had been excluded from enjoyment of "nature's bounty." Abolition-democracy, however imperfect, provided a powerful glimpse of freedom and equality, one in which the ability to experience and use nature figured prominently:

> This was the coming of the Lord. . . . It was everything miraculous and perfect and promising. For the first time in their life, they could travel; they could see; they could change the dead level of their labor; they could talk to friends and sit at sundown and in moonlight, listening and imparting wondertales. They could hunt in the swamps, and fish in the rivers. And above all, they could stand up and assert themselves. (Du Bois 1992, 122)

The promise of abolition-democracy, however, was undercut by a revanchist cross-class alliance forged through the institution of whiteness. In the aftermath of the war, a *dictatorship of capital* quickly emerged in the North (Du Bois 1992, 239), and the working-class segments of the west took comfort in petty bourgeois dreams of becoming cattle barons and oil tycoons:

> An empire of rich land . . . had been snatched from the hands of prospective peasant farmers and given to investors and land speculators. All of the national treasure of coal, oil, copper, gold and iron had been given away for a song to be made the monopolized basis of private fortunes and perpetual power to tax labor for the right to live and work. (Du Bois 1992, 581)

The precarious alliance that constituted the abolition-democracy began to crumble as "property in the south got the ear of property in the north," and poor Southerners opted to become "co-workers with the white planters" by joining organizations like the Ku Klux Klan (Du Bois 1992, 131, 581). The short-lived success of abolition-democracy on the Sea Islands, Du Bois notes, faded into injustice, as "the government broke its implied promise and drove [the formerly enslaved peoples] off the land" (1992, 393). When Andrew Johnson returned property to its antebellum owners, the African American landowners protested to no avail: "What is the use of giving us freedom if we can't stay where we were raised and own our own house where we were born and our own piece of ground?" (Du Bois 1992, 602). Historian Chad Montrie notes that "by the late 1870s and 1880s, the 'redeemed' white elite began passing enclosure or 'fence' laws or more strictly enforcing existing laws against common use of unfenced land, not coincidentally starting in counties with high black populations" (2008, 49). The fence laws were part of a broader enclosure movement, led by the plantation class, that deployed conservationist concerns over scientific resource management in order to extinguish usufructuary rights to hunting, fishing, and grazing (Hahn 1982). While such measures imposed hardship on poor whites as well, their primary aim was to discipline newly freed Blacks whose ability to access nature for their subsistence needs vacillated against their willingness to abide the miserable conditions of plantation labor.[3] The resurgence of a cross-class alliance among whites was the proverbial knife in the back of the abolition-democracy. The allure of access to and control over nature—which workers pined for and elites possessed—was integral to the reconsolidation of whiteness.

> The chief obstacle in this rich realm of the United States, endowed with every natural resource and with the abilities of a hundred different peoples—the chief and only obstacle in the coming of that kingdom of economic equality which is the only logical end of work is the determination of the white world to keep the black world poor and themselves rich. (Du Bois 1992, 706)

CONCLUSION: LOOKING BACKWARD AND LOOKING FORWARD

The *natural wages of whiteness* worked to splinter working-class struggles for fair and equitable wages (in the expansive sense that Du Bois uses the term) in three distinct ways. First, the plentiful nature of America supposedly vacillated against inequality, but white men possessed the sole *right to work the wilderness*: the exclusive legal authority to transform the land and to thereby claim possession of it. Second, the working class was, *naturally*, viewed as the domain of white workers; nature was deployed, in different ways, to subordinate working-class women, people of color, Native Americans, and immigrants. In some cases, particularly in the west and South, white workers did gain from this better access to land and resources—an *environmental wage* that enabled access to sources of subsistence and/or financial gain[4]—but the broader impact was to provide ideological cover for capital that worked to further divide working-class movements and enable the elite capture of nature. Third, abolition-democracy provided one pathway forward, a moment in time where the conditions of possibility existed for cross-racial working-class solidarity predicated, in part, on claims that we would today recognize as environmental: access to and control over land, water, and forests, and the ability to hunt, fish, and recreate in nature.

The downfall of abolition-democracy coincided with the rise of conservationism. By the mid-to-late nineteenth century, looming shortages of fish and timber, concerns over soil fertility, and the decimation of beaver, buffalo, and several bird species had made it clear that nature was not, in fact, infinite, and that corrective action would need to be taken to sustain resource yields into the future (Hough 1873; Marsh 2003 [1864]). This often took the form of land reserved for "public use." A decade-plus after the Miwok had been violently pushed out of Yosemite Valley—part of an effort to open up space for mining interests amid the gold rush—President Lincoln set aside Yosemite through an 1864 land grant to the state of California, with the condition that the land be forever reserved for the "public use, resort, and recreation." Soon after President Grant signed the Yellowstone Park Protection Act in 1872, park administrators, backed by the influential Boone and Crockett Club, used the army to restrict the access of members of Shoshone, Crow, Nez Perce, and other nearby tribes, on the grounds of supposed overhunting by natives—a "scientific" claim belied by the fact that the US Army was waging an intensified onslaught against Native American sovereignty, in part via the destruction of the buffalo on which Plains tribes subsisted (White 1991; Spence 1999; Germic 2001; Jacoby 2001; Estes 2019). Many of the staunchest early conservation policies—for example, the establishment of forest reserves cordoned off from productive use—were opposed by most logging, mining, and grazing interests (Graf 1990, 62–63; Vaughn Switzer 1997, 27), but the conservationist mode of environmental protection posed little threat to a burgeoning ruling-class ecology. To the contrary, it served to expand elite access to "wild" spaces, further Native American dispossession, restrict working-class subsistence practices, and stabilize the conditions of production on which continued capital accumulation depended.

By the 1870s, as the movement to protect wilderness gained momentum amid the closing of the "wild frontier," a new frontier was also opening up. Closer to home, but outside of the cities and their industrial rhythms, "the merchant princes and millionaires were searching for hilltops, shore lands, and farms on which to build substantial estates; crowded cities offered fewer attractions with every passing year" (Jackson 1987, 25). Seeking to reconnect with Nature, they turned to the suburbs.

2 DEFENDING THE SUBURBAN ENVIRONMENT: THE TREADMILL OF SOCIAL REPRODUCTION AND THE MAKING OF THE MIDDLE CLASS

If you seek the monuments of the bourgeoisie, go to the suburbs and look around.
—Robert Fishman, *Bourgeois Utopias*

On May 24 and 25, 1965, amid a growing demand for a stronger federal response to environmental degradation, United States President Lyndon B. Johnson convened the White House Conference on Natural Beauty. While the official title emphasized the aesthetic dimension of environmental protection, the conference comprised fifteen panels covering a variety of conservation-related issues. One panel, "The New Suburbia," was presided over by FitzGerald Bemiss, then chairman of the Virginia Outdoor Recreation Study Commission, a state-sponsored organization that has since been credited with launching Virginia's environmental movement (Peters 2008). A state senator who grew up on Richmond's posh Monument Avenue—where from his home's windows he "could look eyeball to eyeball with [the statue of General Robert E.] Lee's Traveler" (Slipek 1980; see also Hass 2013)—Bemiss became active in conservationism during the 1960s and was lead author of a 1965 report, "Virginia's Common Wealth: A Study of Virginia's Outdoor Recreation Resources," that would prove instrumental in spurring the state legislature to take action to regulate pollution and preserve open space. The commission's "Virginia Outdoor Plan" articulated an early vision of environmental citizenship—to be a responsible Virginian meant to safeguard the state's natural heritage: clean air and water, wildlife, and open spaces.[1]

Bemiss had also, in previous decades, been deeply involved in debates over the desegregation of Virginia schools.[2] In fact, he once wrote a letter to US Senator Prescott Bush encouraging him to read James Kilpatrick's noxious editorials warning of the grave and imminent dangers that racially integrated schools would bring to Southern society (Grundman 1972, 89–90).[3] Considered a moderate in Virginia politics, Bemiss was nonetheless persuaded against "interposition," the neo-Confederate framing of federal intervention as unconstitutional, that was rekindled in the late 1940s and 1950s as the nation seemed poised to confer civil rights on to African Americans.[4] His "centrist" position on race relations was formed, at least in part, through dialogue with his one-time neighbor, Lewis F. Powell, then chairman of the Richmond school board and future Supreme Court justice (Schapiro 2011). The compromise position crafted by Bemiss and other moderates of the day favored "local choice"—that is, school districts could decide for themselves whether to comply with federal integration policy (and desegregate their schools) or to instead close their public schools entirely. Threading the needle between explicit rejection of federal civil rights laws and actually granting access to African American students, local choice would eventually be deployed by many suburbs across the country "as a 'race-neutral' formula that would minimize integration through reliance upon residential segregation" (Lassiter 2006, 14; see also Gates 1964; Grundman 1972; Bartley 1997, 320–345; Smith 1998; Kruse 2005).

The connection between Bemiss's de facto support for school segregation and his conservationism is not a product of ideological inconsistency or individual idiosyncrasy but of a widespread sense of alarm at the rapid changes threatening traditional Southern life. "The face of Virginia," he wrote, "is taking on a new character as it becomes urbanized and industrialized in its commitment to progress" (Bemiss et al. 1965, 4; see also Bemiss 1960). In this regard, Bemiss's politics offer a snapshot into the broader dynamics of the time: a member of the Southern gentry from the former capital of the Confederacy who turned from race to the environment in an effort to articulate a vision of the public interest amid a backdrop of cascading changes. Both narratives—the racial and environmental—were taken up by the white suburban masses in defense of their own perceived interests, serving to conserve a social order that benefited the elite while masquerading as middle-class politics for the common good. The example of Bemiss helps to illuminate the tensions and contradictions that characterized the middle-class ecological order of the mid-twentieth century, and allows us to begin understanding how the suburbs became bastions of both a new conservatism and a new conservationism that took off in the 1960s, indelibly marking US politics in the process.

This chapter proceeds in two parts. First, I seek to shed light on the changing contours of, and linkages between, environmental degradation and class relations during the decades following World War II by turning to *the treadmill of production* (henceforth abbreviated ToP). Scholars of the ToP have documented how new chemical- and energy-intensive forms of production both destabilized ecosystems and weakened labor movements between the 1940s and the 1970s in crucial ways that reverberate into the present. By putting this scholarship into conversation with social and ecological analyses of the period, we can glean insight into the concept's utility for understanding this historical moment—where a widespread but uneven affluence undercut the power of organized labor and brought ecosystems to the brink of collapse. Present, but perhaps undertheorized, in this body of work is attention to the ways in which shifts in production also transformed life outside of the factory, unleashing a host of socioeconomic pressures and environmental burdens that contributed to the production of the prototypical middle-class environment: the suburbs. In an attempt to understand the complex and, at times, counterintuitive politics that emerged in the suburbs, I engage with socialist feminism and ecofeminism in introducing the concept of *the treadmill of social reproduction*, which emphasizes the growing burdens placed on "care work" and the shifting socio-ecological relations within which this work took place.

Next, I attempt to develop this concept through an analysis of primary and secondary sources on mid-twentieth-century US suburbanization. As rapid industrialization led to higher wages and a bevy of consumer goods, it also produced ecological degradation that a growing swathe of Americans sought to avoid; the rush to the suburbs was a product of the widespread desire of upwardly mobile whites to escape the perils of a racialized urban environment (Jackson 1985; Davis 1986; Sugrue 1996; Lipsitz 1998; Kruse 2005) and to reconnect with nature (Hays 1987; Sale 1993; Rome 2001; Sellers 2012). By the 1960s, however, the suburban environment was widely viewed as in crisis, and suburbia's residents were catapulted onto the front lines of the political interventions that followed. In a regime of production predicated on a "family wage" for male breadwinners, the overwhelmingly white, middle-class women of the suburbs found themselves responsible for keeping up with the demands of an accelerating treadmill of social reproduction, which increasingly extended beyond the home and into political duties related to community stability—for example, membership on school boards, homeowners' associations, neighborhood protection organizations, and local anti-sprawl and conservation groups (Sugrue 1996; McGirr 2001; Geismer 2015). In different ways, but using strikingly similar legal tools, like zoning regulations, such organizations worked to defend the suburban home and neighborhood from disturbances to the environment—both cultural and ecological. These local movements came to intersect with national political forces in galvanizing, on the one hand, a suburban environmentalism in which open space and clean air and water were safeguarded as consumer amenities for middle-class property owners and, on the other hand, a New Right in which reconfigurations of property rights via building and land-use codes became the go-to mechanism for keeping out "undesirables" amid calls for civil rights and economic equality.

Beyond the organizational forms that emanated out of suburbia, the political implications of the "suburban environment" were—and continue to be—far-reaching. The suburban environment was not only a pull-factor that drew white- and blue-collar workers out of industrial centers, it also became a normative standard in relation to which an ostensibly obsolete working class and an emerging "underclass" became intelligible. As the white middle-class habitat par excellence, the suburban environment came to function as an ideological trope that dulled working-class consciousness—the affluent and acquiescent alternative to the palpable anger of an Other America. Whereas the wild frontier furnished the material and ideological ballast for the nineteenth-century wages of whiteness, the suburban environment was one of the reefs on which dreams of twentieth-century US social democracy were shipwrecked. Examining the relationship between the two treadmills, of production and social reproduction, can offer insights into strategies for rekindling working-class politics at a moment when the suburban environment retains much of the political potency that it acquired in this bygone era.

THE TREADMILL OF PRODUCTION

In the years after World War II, the enormous productive capacity that the United States brought to bear on the war effort found new outlets in mass production for an increasingly affluent, consumer society. The "national output of goods and services doubled between 1946 and 1956 and would double again by 1970" (Cohen 2004, 121). Housing starts increased from 142,000 in 1944 to nearly 2,000,000 in 1950 (Cohen 2004, 197). New car sales skyrocketed from 2.1 million in 1946 to 7.9 million in 1955 (Wells 2012, 279). From 1946 to 1951, "consumer spending increased 60 percent, but the amount spent on household furnishings and appliances rose 240 percent" (May 2017, 157). These new houses, cars, and appliances were built from resources withdrawn from nature, often with the aid of new chemicals manufactured by humans, all on a foundation of fossil fuels. The environmental impacts of this growth in production, the shift in the inputs required for production, and the concomitant production of more and different forms of pollution, were quite literally earth-changing.

As environmental consciousness and activism increased over the course of the 1960s and early 1970s—culminating in a broad range of federal laws, including the Wilderness Act, Endangered Species Act, National Environmental Policy Act, Clean Air Act, and Clean Water Act—scholars set out to study the reasons behind this postwar environmental transformation and its impacts on society. Arguably the most consequential concept to have emerged out of this initial wave of environmental social science was the "treadmill of production," first articulated by Allan Schnaiberg (1980), and then popularized by Schnaiberg and a group of students and collaborators, including Kenneth Gould, David Pellow, and Adam Weinberg (see, e.g., Gould, Schnaiberg, and Weinberg 1996; Schnaiberg, Pellow, and Weinberg 2002; Gould, Pellow, and Schnaiberg 2004, 2008). The body of work employing the treadmill of production is substantial, but the basic ideas of the concept can be summarized in three parts: (1) the shift to a Fordist mode of production gave rise to a growing contradiction between the requirements of ecosystems and industrial capitalist economies; (2) the treadmill of production grew into a hegemonic institution, embraced by capital, the state, labor unions, and even many environmentalists; and (3) the treadmill of production created new material and ideological barriers to the success of both labor unions and environmentalists, and to coalitions between the two social forces. These obstacles could only be surmounted—the treadmill could be slowed—through collective action leading to the redistribution of risk and power.

First, Schnaiberg argued that the environmental crises of the time were best explained not by turning attention to consumption, population growth, or technology, but by focusing on transformations in relations of production. The basic contours of the ToP analysis are Marxist in orientation: the post–World War II era witnessed a shift in production away from *variable* and toward *constant* capital; put differently, surplus value was extracted from labor and then invested in new mechanized technologies that diminished the need for that labor. Gabriel Kolko reported, for example, that "the CIO Oil Workers, surveying the impact of automation on the industry, dolefully reported that while refinery production rose 22 percent from 1948 to April 1954, the number of production workers had fallen from 147,000 to 137,000" (1955, 369). In her analysis of the mid-twentieth-century steel industry, Judith Stein found that "between 1937 and 1960 blue-collar employment in steel fell by 10 percent, despite the doubling of production" (1998, 25). In 1963, autoworker and organic intellectual James Boggs noted that "the machine shop work which had been done by 1,800 at the old Chrysler-Jefferson plant was now being done by 596 in the new Trenton, Michigan plant which supplies not only the old plant with machined parts but all the other plants of the corporation" (Boggs 1963, chap. 1). In addition to money saved on union wages, mechanization also allowed capital to avoid the quotidian power struggles that typified industrial politics. "Automatic equipment," wrote Kolko, "does not strike, talk back to foremen, file grievances or do any of the things presently obnoxious to management" (Kolko 1955).

The novel facet of Schnaiberg's argument, where he added a distinct layer to Marxist accounts of the period, was his recognition that this shift to capital intensivity, subsidized in myriad ways by the state, was also both chemical and energy intensive. Postwar production was dependent on *withdrawals* from ecosystems (in the form of water, oil, coal, minerals, timber, etc.) and *additions* to ecosystems (in the form of industrial pollution) that differed, quantitatively and qualitatively, from the withdrawals and additions characterizing previous forms of production. Over the course of the 1950s, for example, water withdrawals in the United States rose from 180 billion gallons a day to 270 billion (Dieter et al. 2018); crude oil production increased from roughly 5 million barrels a day to 7 million (US Energy Information Administration 2021); synthetic organic chemical production nearly doubled, from 49 million pounds to 97 million pounds (US Tariff Commission 1952, 1962); and uranium concentrate production exploded from .36 million pounds to 35 million pounds—in addition to roughly 70 million pounds that were now being imported (US Energy Information Administration 2020). Overall, this emergent mode of production required more withdrawals and produced more pollution that persisted for longer periods of time and, in many cases, bioaccumulated in plants, animals, and humans alike. Insecticides, fungicides, and defoliants originally produced for the war effort were quickly oriented toward domestic agriculture; toxic metals, like chromium, cadmium, and lead, were used in greater quantities and put to work in different forms; synthetic chemicals, like polychlorinated biphenyls and per- and poly-fluoroalkyl substances (PFAS), were devised to produce plastics, fabrics, and nonstick coatings; and radioactive elements like uranium were extracted for Cold War military-industrial expansion.

The result was a tension between what Schnaiberg termed "ecological production" and "social production"; the long-term requirements for eco-systemic stability versus the requirements for economic growth; a tension that lay in the "contradictory utilization of surplus in the two systems." Ecosystems take surplus energy and permit "the growth of just enough species and populations to offset the surplus," but industrial capitalism reinvests surplus into new technologies that substitute "higher levels of energy and a wider array of chemicals" for human labor (Gould, Pellow, and Schnaiberg 2008). Ecosystems become more complex over time and their growth slows. Industrial economies also become more complex, but are hitched to a growth machine driven by the imperatives of capital accumulation (Schnaiberg 1980, 19). The treadmill thus accelerates, "producing yet fewer worker benefits at a given rate of natural resource extraction" (Gould, Pellow and Schnaiberg 2008, 14).

In spite of this glaring contradiction between the long-term interests of capitalists, on one side, and those of workers and ecosystems, on the other, ToP scholars suggest that there existed enormous social buy-in to the treadmill of production: among capitalists, the state, workers, and even many environmentalists. That capitalists supported it is no surprise, as it clearly worked in their interests—there were profits to be made and shareholder value to be maximized. The state, while a multifaceted set of institutions and actors, also acted as if its general interests were aligned with the expansion and acceleration of the treadmill: from tax policy that underwrote investment in new factories and technologies to massive subsidies for oil and gas production to the adoption of a laissez-faire attitude toward the risks of new synthetic chemicals (Schnaiberg 1980, 131; see also Pellow 2000).

ToP theorists also point out that workers had enormous material incentives to see their interests as aligned with the growth machine. According to Gould, Schnaiberg, and Weinberg, "Workers believed that increased production created new employment opportunities both in direct industrial production and, more indirectly, in the construction and service sectors" (1996, 5–6). Insofar as a portion of growth in profits and production, however asymmetric, was shared with workers, this belief was not mere ideology: in the postwar world, wages generally grew alongside production. As Marxist sociologist Michael Burawoy has argued, "Through the concessions and high living standards associated with an advanced capitalist economy, the interests of capital and labor [were] concretely coordinated" (1985, 28). The sway that the treadmill of production had on labor was strengthened in the 1960s and early 1970s when capital flight to the "right to work" South was in full swing and companies began to loudly proclaim that "costly" and "burdensome" environmental regulations were major sources behind the job cuts to come.

This social buy-in to the treadmill was, of course, not uniform; the period was punctuated by waves of discontent both inside and outside the factory.[5] And yet, the pull and pace of the mid-twentieth-century treadmill of production was astounding. The treadmill not only dulled the power and vision of labor unions, it also worked to sever potential connections between unions and those fighting against environmental degradation. Schnaiberg suggested that the shift to capital-intensive production ironically rendered the exploitation of nature more difficult to see (1980, 131–133). By the early 1970s, the fact that there were, in general, fewer people employed in direct production and extractive industries helped to stimulate an emerging societal common sense, which held that we were rapidly entering a "postindustrial era." This line of thinking, while attuned to overarching changes in the contours of production, ignored the growing withdrawals and additions that were still required to feed the industrial beast, and rendered invisible the ways in which consumers were able to distance themselves from the ecological harms created by an increasingly *global* treadmill of production. American industry required growing quantities of "steel and iron for metallic frames, copper for wiring, and petroleum for plastics and (most significantly) fuel" (Black 2018, 122). In addition to coal and uranium extracted on Native American lands, these primary products were increasingly appropriated from abroad, redistributing both jobs and environmental burdens across borders.

While labor unions generally supported early antipollution measures (see, e.g., Dewey 1998; Gordon 1998; Obach 2004; Montrie 2008; Loomis 2016; Rector 2018; Hultgren and Stevis 2020; Rector 2022; Stevis 2023), environmentalists came to emphasize protections—like wilderness areas, wild and scenic rivers, and open spaces—in the spheres of residential life and leisure that many workers begrudgingly rejected as opposed to and less pressing than their need for a steady paycheck. Such measures laudably cordoned off particular spaces from commodification but paid too little attention to the distribution of environmental amenities/harms and ultimately failed to slow the treadmill in any meaningful way. This division between environmentalists and unions was aggravated by the decomposition of the industrial working class, a process taking place within and outside the factory gates. In a workplace undergoing automation and capital flight, some workers found themselves permanently unemployed or underemployed, while others were integrated into middle-management or were "incorporated into the new physical (and electronic) technological systems" (Gould, Pellow, and Schnaiberg, 2004, 299). At home, already existing patterns of urban housing segregation on the basis of race and class were being amplified by suburbanization:

> The middle classes lived upwind and upstream from polluting enterprises. Blue-collar workers were induced and/or coerced to live downwind or downstream or adjacent to polluted communities because of their lower property costs and the limited wages of the workers. (Gould, Pellow, and Schnaiberg 2008, 13)

This segregation further divided workers, and drove a spatial wedge between organized labor and organized environmentalism. The end result: an expansion of the treadmill, leading to more mechanization and fewer workers, while creating "continuous ecosystem disorganization, at minimum through the destruction of land and habitats and, more typically, involving further pollution and depletion of natural resources" (Gould, Schnaiberg and Weinberg, 1996, 7). Neither workers nor environmentalists were happy about this reality but consumer goods dulled the pain for all, at least temporarily, and residential mobility offered a means for some to detach home life from the devastating social and environmental effects of late-industrial capitalism. "Middle class workers, who benefited from the expansion of the treadmill, largely moved to emerging bedroom suburbs," write Gould, Pellow and Schnaiberg (2008, 13). How might what happened within these bedroom suburbs relate to the social and ecological tendencies discussed here? The discussion that follows elaborates on the concept of social reproduction and considers its relevance to analyses of mid-twentieth-century suburban politics.

FROM TREADMILL OF PRODUCTION TO SOCIAL REPRODUCTION

There is no doubt that "the process of production decisively shapes the development of working-class struggles" (Burawoy 1985, 7), but capital accumulation also depends on "a background of non-commodities," which include nature, state power, and unwaged labor (Fraser 2017). Socialist feminists have long argued that "the material base of patriarchy is the work that women do reproducing the species" (Hartmann 1979, 9). The concept of social reproduction extends this emphasis on the role of biological reproduction within capitalism—often traced back to Engels's *Origin of the Family, Private Property and the State* (1884) [6]—to the broader array of non-waged social activities necessary for sustaining human society. Social reproduction refers quite simply to "the everyday activities of maintaining life and reproducing the next generation" (Bakker 2007). As Nancy Fraser argues, "Wage labor could not exist in the absence of housework, child-raising, schooling, affective care and a host of other activities which help to produce new generations of workers and replenish existing ones, as well as to maintain social bonds and shared understandings." These non-marketized social relations provide the conditions of possibility for markets (Fraser 2017, 59–60).

Social reproductive labor (raising children, cleaning, cooking, caring for the sick and elderly, educating the next generation, and so on) was, in the political conjuncture in which Schnaiberg and his colleagues initially focused their efforts, radically gendered, in the sense that women were responsible for the lion's share of it, even as many—particularly working-class women, disproportionately women of color—continued to work outside of the home (Davis 1983, 226–233).[7] Indeed, the ToP emerged to explain a conjuncture wherein the "family wage," in which a "working man's" remuneration was ample enough to support his wife and children, was one of the foremost demands of unions. Although women's participation in the labor force expanded considerably during this era—in ways shot through with racial inequalities—this social demand reinforced a reality in which many women were confined to the "domestic sphere."[8] As I noted in chapter 1, historians have detailed how the distinction between "productive" and "domestic" work was institutionalized during the mid-nineteenth century, and was dependent on the reduction of unwaged labor to a natural dimension of women's very being. Silvia Federici would later remark that housework has been "transformed into a natural attribute of our female physique and personality, an internal need, an aspiration, supposedly coming from the depth of our female character. . . . Capital had to convince us that it is a natural, unavoidable and even fulfilling activity to make us accept our unwaged work" (1974, 77).

The labor of social reproduction is bound up in ecological processes in ways that differ from, but are intricately connected to, the dynamics outlined by the treadmill of production. In a 1987 essay, titled "The Theoretical Structure of Ecological Revolutions," Carolyn Merchant argued that "ecological revolutions" occur during periods of shifting interactions among *production* (the "extraction, processing and exchange of nature's parts as resources"), *reproduction* (the ways in which humans are "born, naturalized, socialized and governed"), and *ecology* (the material requirements of ecosystems, and dominant human ideas and representations of nonhuman life). These foundational elements of our lives are dialectically entwined—in order to productively work outside the home, our basic material needs have to be met within it; and in order to meet these needs, we rely on stable sources of food, water, and natural resources. As eco-feminists have long argued, social reproductive labor has often centered around growing, procuring, and preparing sources of sustenance, such as water and firewood from village commons, and fruits and vegetables from family and community garden plots (Salleh 2005; Mellor 2006; Mies and Shiva 2014). With such practices in mind, Ariel Salleh argues that social reproductive labor—"women's historically assigned caring and maintenance chores"—has functioned as a bridge connecting men (historically assigned waged labor roles) to nature (2003, 67). However, the progressive encroachment of industrial capitalism into everyday life has transformed social reproductive labor and its linkage to nature. By the mid-twentieth century, suburban women's relationship to nature was thoroughly mediated by the treadmill of production. Food was procured through trips to grocery stores in suburban strip malls, meal preparation increasingly relied on processed and frozen foods, and household cleaning was done with the aid of synthetic-chemical sprays and detergents. Not only was access to the means of subsistence dependent on the waged labor earned by men, but the social reproductive labor historically anchored in direct contact with nature was now channeled through the automobile, pesticides, and newly invented chemicals. A rift had opened up into the metabolic relationship between nature and social reproductive labor.

There exist two treadmills—of production and social reproduction—that are intimately entwined. The ongoing expansion and acceleration of the ToP, alongside the gender dynamics, racial inequalities and geopolitical anxieties that characterized industrial life in the 1940s and 1950s, affected the terrain of social reproduction. Even with new time-saving technologies and consumer goods (washing machines, vacuum cleaners, frozen meals, etc.), the care work necessary to maintain a suburban household and to stabilize the immediate social and ecological surroundings—what we might think of as the *treadmill of social reproduction*—had expanded and accelerated.[9] With their husbands working long shifts and commuting to and from the city, the overwhelmingly white, middle-class women of suburbia found themselves with more work to do—more kids to have and care for and bigger houses to clean, but also new duties that were *at once outside of the household but related to its vitality and stability*: organizational forms pertaining to the education of children, the preservation of parks and open spaces, the cleanliness of local air and water supplies, and the racial, class, and ideological composition of the neighborhood and local community. This political work was experienced as both a much-needed outlet for participation in the polis—a way to ameliorate the "nameless, aching dissatisfaction" of which Betty Friedan famously spoke—and an additional, uncompensated form of labor that was nonetheless integral to the stabilization of social life (Hurley 1995, 556–557).

It is on the terrain of social reproduction—peoples' stubborn desire to breathe clean air, have time with families, live within a stable environment, and so on—that the treadmill of production has met some of its most impactful resistance. It is also on the terrain of social reproduction, however, that the emancipatory promise of resistance has so often fallen victim to the powerful "wages of whiteness" (Du Bois 1992) discussed in chapter 1. As the crises created by racial capitalism spread and began to threaten the habitability of environs outside of industrial centers, the residents of mid-twentieth-century suburbia made demands primarily through local political mobilizations—reactionary, progressive, and often somewhere in between—that emerged out of a complex interplay between dissatisfaction with and threats to the stability of the suburban social reproductive structure. In a suburban environment plagued by destabilizing forces originating from the world beyond, suburban women, in particular, acquired a political voice as representatives of the family, guardians of domesticity, and defenders of the local environment.

CONSTRUCTING THE SUBURBAN ENVIRONMENT

Historian Lizabeth Cohen writes that "between 1947 and 1953, the suburban population increased by 43 percent, in contrast to a general population increase of only 11 percent; over the course of the 1950s, in the twenty largest metropolitan areas, cities would grow by only .1 percent, their suburbs by an explosive 45 percent" (2004, 195). Whereas, on the brink of World War II, only 44 percent of Americans owned their own homes, by 1960, rates of homeownership had risen to 62 percent (Cohen 2004, 123). "Nearly all of the increase," according to historian Christopher Wells, "came in car-dependent suburbs, where homeownership nearly doubled between 1940 and 1950" (2012, 255; see also Jackson 1985, 191). These new suburban homes were owned overwhelmingly by white Americans. "One of the greatest migrations of the twentieth century," wrote Thomas Sugrue, "was the movement of whites from central cities to suburbs." According to the 1968 Kerner Commission Report, "From 1950 to 1966, 77.8% of the white population increase of 35.6 million took place in the suburbs. Central cities received only 2.5 percent of this total white increase" (1996, 118).

There were, of course, several other migratory flows occurring over the first half of the twentieth century that preceded and precipitated the rush to suburbia. "From 1880 until 1924 . . . almost 25 million people came to the United States from other nations," the largest share from southern, central, and eastern Europe (Rome 2008, 433). Not yet welcomed into the sphere of whiteness, these immigrants—from Italy, Poland, Czechoslovakia, Lithuania, Russia, and so on—disproportionately worked dirty and dangerous jobs (e.g., as janitors, junk and scrap traders, garbage collectors), lived in slums plagued by industrial pollution and a lack of sanitation services, and faced sustained attacks from eugenicists and nativists for their purportedly uncivilized relationships with nature (Washington 2005, 75–97; Rome 2008; Zimring 2017, chap. 5). Due to eugenics-era immigration restrictions implemented in 1917 and 1924, immigration from Southern and Eastern Europe, as well as Asia, ground to a near halt, and capital came to rely on migrants from the US South (both Black and white) and from Mexico. Over the first half of the twentieth century, nearly five million white Southerners migrated to regions other than the South, settling in rural and urban areas in the North and west (Gregory 2005, 14–17; Bazzi et al. 2023, 5–6). But the migration that influenced the process of suburbanization most directly was that of African Americans fleeing the realities of Jim Crow.

Beginning around 1910 and then accelerating during World War I, the Great Depression, and World War II, African Americans streamed into industrial centers of New York, Chicago, Philadelphia, Denver, Detroit, Kansas City, and beyond. As the mechanization of agriculture intensified alongside continued racist violence, over 1.4 million Blacks left the South during the

1940s, 1.1 million in the 1950s, and 2.4 million in the 1960s (Gregory 2005, 14). Between 1940 and 1950, the number of African Americans outside the South "leaped more than 50 percent, from 3,986,000 to 5,989,000." By 1960, this number had climbed to 9,000,000 (Lubell 1965, 98). By this point, "a majority of America's African American population lived in cities, most of them north of the Mason-Dixon line" (Sugrue 1996, 7; see also Jackson 1985; Gregory 2005; Wilkerson 2010).

The migrations of African Americans to the North and of the white middle class to the suburbs were intimately entwined, a fact that was absent from many of the popular depictions of the time, which tended to naturalize the process of suburbanization and depoliticize the everyday realities of suburban life. Both the news media and popular literature generally portrayed the suburbs as places where middle-class residents led intense social lives but avoided ideology like the plague. Popular sentiment held that the social crowded out the political in suburbia; it was all *kaffeeklatsch* and no partisan clash, stewing anxiety about status without any of the power asymmetries that had historically animated class relations. While some scholars of the era pushed back against such over-simplified caricatures (Wood 1958; Berger 1960), this dominant image of suburbia held enormous sway, not only as a description of suburban life but as a window into the changing nature of the country. The suburbs were, according to sociologist Bennett Berger (1960), "rich with visible symbols readily organizable into an image of a way of life which could be marketed to a non-suburban public" (Berger 1960, xvii). As political scientist Robert C. Wood put it, "The most fashionable definition of suburbia today is that it is a looking glass in which the character, behavior and culture of middle-class America is displayed" (1958, 4).

Contemporary historians have pointed out that the reality was dramatically different from the image: from the suburbs a profound backlash against "socialism" was launched, as well as against movements for racial justice and women's rights; from Orange County, California, to Grosse Pointe, Michigan, to Fairfield, Connecticut—the dredges of McCarthyism, Goldwater fever, and segregationist venom occupied a special place within the "little boxes" of suburban life (Davis 1986; McGirr 2001; Lassiter 2006). The New Right was, in many respects, a revolt from the suburbs (Davis 1986, 163, 176). Suburbs were also, however, places in which the ravages of toxic contamination, sprawl, and the decline of wildlife populations touched the sensitive nerves of those who had fled the industrial centers seeking a better environment (Rome 2001; Sellers 2012). Indeed, the suburbs were central to the stabilization of "the environment," *as a concept*. Born in the trenches of World War I, and codified in the industrial health divisions of factories (Sellers 1997; Bond 2022), the environment—understood as a web of interdependent linkages among human and nonhuman lives and systems—entered into widespread popular consciousness, in large part, through the struggles being waged in the suburbs (Sellers 2012). The discussions that follow detail the ways in which the US suburbs emerged alongside a growing concern for the environment.

The Fall of the Suburban Commons and Rise of the Middle-Class Environment

In his magisterial history of the suburbs, *Crabgrass Frontier*, Kenneth Jackson (1985) notes that, by the mid-1870s, the rich—anxious about the overcrowded and unsanitary conditions of urban centers—were seeking prime real estate beyond municipal boundaries. At the same time, however, there existed a separate trajectory of suburban development, wherein working-class communities found cheap housing, land, and opportunities for subsistence living. In the suburbs of the wealthy, nature was prized for its health effects, its beauty, and its distance from a crowded and contaminated industrial life. In the suburbs of the working class, nature functioned as an "informal commons," where people grew food, raised livestock, and gathered firewood for their own consumption, in some cases selling the excess to fellow suburban residents (Sellers 2012, 17–22). In ideal-typical suburbs, the experience of the country was part and parcel of suburban living. As environmental historian Christopher Sellers details, however, over the course of the 1920s and 1930s, "the suburban country" became a contradiction in terms. Formerly widespread practices of informal commoning were deemed public health threats and outlawed via land-use codes (see also Baker 2018, 18–19). The enclosure of the suburban commons was institutionalized, in part, through early forms of redlining. As Sellers documents, "Places with too many 'sustenance homesteads' could also be classified as poor credit risks" (2012, 35). He continues, citing records from LA County Home Owners' Loan Corporation (HOLC) survey files:

> Most horrifying (and D-rated) of all were not downtowns, but places that were all too rural and backward as well as nonwhite, such as one corner of Los Angeles's rural fringe that was hopelessly abandoned to an "infiltration of goats, rabbits and dark-skinned babies." (2012, 35)

Although the suburban commons were gradually being privatized, the promise of nature—"an acre of land, a garden, flowers"—continued to draw migrants from the city, even as a new ideal-typical suburb emerged: home to the upwardly mobile, white, "middle class." In his history of US environmentalism, Kirkpatrick Sale writes that "the postwar material boom . . . produced a growing number of the college-trained, white-collar middle class that increasingly populated suburbs in search of trees and birdsong and stars at night" (1993, 7). The new suburbanites fled the cities in the quest for a better quality of life: "the amenities beyond the necessities, including leisure time, outdoor recreation, healthy air and water, personal health and security, and hence a greater emphasis on the natural world, at least as represented by parks, preserves, wilderness areas, forest reserves, botanical gardens, and scenic highways" (Sale 1993). Environmental historian Samuel Hays concurs that "the most widespread source of emerging environmental interest was the search for a better life associated with home, community and leisure. . . . Millions of urban Americans desired to live on the fringe of the city where life was less congested, the air cleaner, noise reduced, and there was less concentrated waste from manifold human activities" (1987, 22–23). The cost of the daily commute, according to a 1955 article in the *Saturday Evening Post*, was worth the trade-off for "a backyard pool and fresh air." The pull of suburbia was particularly strong for the prototypical (white middle-class) mid-twentieth-century American husband and father:

> He will have more than five rooms, his children will have fresh air, play space and better schools, the house will be his in an area of rising values, and his family will be a little beyond bomb range of New York—a lurking consideration in the back of more than one man's mind. (Rowland 1957)

While nuclear dangers may have lurked in the background, the promise of peace and prosperity was etched into many analyses of suburban class relations. In the eyes of most commentators, the middle-class suburbs were not sites of class relations at all, but functioned as the household counterpoint to the purportedly amicable industrial arrangements that prevailed in the glow of the newfound US global hegemony. In William Whyte's telling, the "organization men" of suburbia "confound the usual concepts of class. . . . Some can be described as upper class, some middle class, but it is the horizontal grouping in which they come together that is more significant" (1956, 269–270). A 1956 *Saturday Evening Post* article, "They Commute to the Wilderness," echoed this sentiment in its depiction of the residents of Mill Valley, California—a San Francisco suburb where bank executives lived next to carpenters, wild animals were frequent houseguests, and all community members banded together in shared appreciation for their natural surroundings.

The quest for nature certainly did fuel suburbanization, but such a recognition ought not feed into a narrative that sanitizes the social history of the suburbs. The pull of the suburban environment is not reducible to an effort by the white middle class to perpetuate social exclusion, but neither can it be disconnected from the racial prejudice and class dynamics that drove white flight. The social history of racial and class exclusion from suburbia has been written many times over: African Americans in particular, racial minorities in general, and "white ethnics," in some cases, faced systematic and sustained discrimination from the federal government, banks, developers, real estate agents, and white residents. Housing discrimination took multiple forms, from redlining, blockbusting, and predatory lending to racially restrictive covenants, intimidation, and violence at the hands of neighborhood protection associations and their supporters (Sugrue 1996; Kruse 2005; Lassiter 2006; Coates 2014). Dorceta Taylor writes that "between 1935 and 1950, roughly 9 million new homes were constructed by private developers in the U.S. However, minorities gained occupancy to only about 100,000, or 1.1% of these" (2014, 241).[10]

As civil rights advocates successfully tore down the racist barriers that characterized housing, education, jobs, and transportation, the retreat of whites to the suburbs grew. In *White Flight*, historian Kevin Kruse notes that while national politicians waged a reactionary struggle in the courts and Congress to preserve the old system of *de jure* segregation, those at the local level were discovering a number of ways in which they could preserve and, indeed, perfect the realities of racial segregation outside the realm of law and politics. Ultimately, the mass migration of whites from cities to the suburbs proved to be the most successful segregationist response to the moral demands of the civil rights movement and the legal authority of the courts. (2005, 8)

In the suburbs, middle-class whites were able to insulate themselves from perceived social transgressions not by turning to the legal apparatus of Jim Crow, but through the more quotidian local matters of school choice and neighborhood zoning. After redlining and racially restrictive covenants were de jure outlawed, new forms of less-overt residential discrimination took hold. Zoning—related to land use, architectural standards, single-family occupancy, and lot size—was at the forefront of these exclusionary efforts (Sugrue 1996, 45; Sellers 2012, 56). One common form of such discrimination was the minimum-acreage requirement. Integral to a broader set of instruments that can be thought of as *ecologically restrictive covenants*, the stipulation that any house must be sited on a minimum amount of land (often one or two acres) worked to provide an ostensibly class- and race-neutral language for ensuring that the status and complexion of the community remained homogeneous. The "two-acre aristocracy" (Burton 1955) was the most common such restriction, but this was complemented by density limits (e.g., restrictions on multifamily residential housing), and eventually eclipsed by efforts to restrict growth-writ-large by invoking concerns over water supply, sewage capacity, or open space. If the suburbs represented the "institutionalization of mobility" (Newman 1957), this mobility was largely confined to white Americans. The ostensible classlessness of suburbia, Whyte wrote, "stops very sharply at the color line" (Whyte 1956, 311).[11]

Defenders of the Middle-Class Environment: The New Right

By the 1960s, the idyllic environment that had drawn so many Americans to the suburbs was widely perceived to be contaminated; the problems of the cities, both social and ecological, were rapidly encroaching on the suburban dream. "Life in the suburbs, wrote the architect and critic Peter Blake, "has become only a little less intolerable than life on congested tenement streets" (1963). Suburbs were far from monolithic, but suburban development followed a general trajectory that influenced the resulting political mobilizations: the deteriorating conditions of urban centers—coupled with efforts to integrate social life and institutionalize civil rights protections for African-Americans and other racial minorities—resulted in massive (overwhelmingly white) flight to the suburbs; suburban privilege was protected both by spatial distance and staunch local organization; the movements emanating out of suburbia eventually adopted a race- and class-neutral language that sanctified private property and "freedom of association"; and the institutionalization of the suburban environment worked to spatialize racial categories and expand full-fledged whiteness to so-called white ethnics—what Nell Irvin Painter calls "the third enlargement of whiteness" (2010, 365–373).[12] While residential segregation on the basis of race was far from new (Massey and Denton 1993),[13] postwar suburbanization served to further entrench racial categories by cementing them in the day-to-day lives of many residents: "Blackness and whiteness assumed a spatial definition. . . . The completeness of racial segregation made ghettoization seem an inevitable, natural consequence of profound racial differences" (Sugrue 1996, 9).

The reactionary dynamics of suburbia nonetheless took on regionally specific forms. In the South, suburbanization occurred amid the backdrop of "massive resistance" to civil rights movements and attempts at desegregation, which were viewed by adherents of the status quo, like Mississippi Senator James Eastland, as an "attempt to graft into the organic law of the land the teachings, preachments, and social doctrines arising from a political philosophy which . . . can be traced to Karl Marx" (Bloom 1987, 95; citing Eastland 1955; see also Bartley 1997, 118–122). While the Calhounian doctrine of interposition attracted enormous attention from neo-confederate advocates of state's rights, at a grassroots level many Southerners framed their supposed plight as a struggle to maintain individual rights: "the 'right to select their neighbors, their employees, and their children's classmates, the 'right to do as they pleased with their private property and personal businesses, and perhaps most important, the 'right' to remain free from what they saw as dangerous encroachments by the federal government" (Kruse 2005, 9). White men were the most prominent opponents of desegregation, but white women also played vital roles within the struggle to save Jim Crow, functioning as what Elizabeth Gillespie McRae terms "segregation's constant gardeners":

> In a Jim Crow nation, segregation's female activists imbued women's civic duties, womanhood, and motherhood with particular racist prescriptions. For many, being a good white mother or a good white woman meant teaching and enforcing racial distance in their homes and in the larger public sphere. (2018, 4)

In the suburbs of the Sunbelt, the debate centered less explicitly around the re-entrenchment of Jim Crow and more around the specter of communism. In *Suburban Warriors* (2001), Lisa McGirr details how the planned sprawl of Orange County, California, created the conditions under which communal ties were severed and anticommunism emerged as a common link among suburban residents as well as between seemingly disparate political movements, like libertarianism and evangelical Christianity. While men disproportionately occupied the leadership positions of the organizations advancing such a narrative, suburban women bought the message and did the bulk of the painstaking work of political organizing:

> According to a study of the John Birch society in California during the 1960s, women were in fact overrepresented among the rank and file of the movement. . . . Not only were women activists themselves, it was also not unusual for women to first become involved and then bring their husbands into the cause. . . . They also enlisted female neighbors and friends and developed cooperative networks so that young mothers could participate in conservative activity. (McGirr 2001, 87)

This political activity did not remain long within the privatized home, but burst forth into the interstices of suburban social life: "Bridge clubs, coffee klatches, and barbeques—all popular in the new suburban communities—provided some of the opportunities for right-wing ideas to spread literally from home to home throughout the county" (McGirr 2001, 97).

In her study of Los Angeles County, for example, Michelle Nickerson illustrates how a sizeable network of Republican Women's Clubs met in the homes of local activists, cultivating both social ties and political outreach. The outreach spread from house to house across suburbs and into larger venues in city centers: "The region's expanding freeway system and sprouting residential centers enabled middle-class women—owners of cars—to shuttle between political duties in the suburbs and events downtown" (2012, 33). Drawing on their socially accepted roles as nurturers and managers of the domestic sphere, suburban female activists confronted progressive education, pediatric mental health experts, supposed communist infiltration of school boards and PTAs, perceived local manifestations of United Nations internationalism, and racial integration of neighborhoods and schools—all linked by an opposition to centralized power and a messianic faith in anticommunism. By defending the home, neighborhood, and local community, conservative women saw themselves as the first-line defenders of the nation, a duty that they pursued with zeal:

> While ink dried on the first issues of the *National Review*, mimeograph machines across Los Angeles County spat out newsletters, many of them composed and printed by teams of housewives. . . . They squeezed meetings, study, writing, and printing into daytime and nighttime hours between trips to the grocery store, meal preparation, and help with homework. (Nickerson 2012, 38)

The "housewife populism" (Nickerson 2012) that emanated from the suburbs, expressed in a discourse of parental and state's rights, had national reach (McRae 2018, 214). The suburbs of the South and Sunbelt were the test kitchens in which ideas that had progressively been building steam in conservative circles—via *Human Events*, *National Review*, the *Dan Smoot Report*, the Bircher newsletter *American Opinion*, and the like—were sampled and shared.

Many of these ideas would make themselves nationally known through the presidential campaign of Barry Goldwater in 1964. Goldwater's campaign provided an opportunity for the Republican Party to appeal to the Jim Crow South using the language and philosophy of libertarianism. Republicans had long been trying to build inroads among Southern whites; the New Deal coalition—a fragile achievement that combined labor unions, African Americans, so-called ethnic whites, and southern Democrats—was ideologically scattered and vulnerable to collapse. Goldwater's eventual victories in the Southern states of Mississippi, Louisiana, Alabama, South Carolina, and

Georgia were—in spite of his otherwise thundering national defeat—cause for continued efforts in this vein. The defeat of Goldwater also chastened the Republican Party, though, and the Nixon campaign (and later administration) faced a delicate effort to balance its multiple constituencies while continuing to pull apart the increasingly precarious balance of the New Deal coalition. While Nixon's vaunted Southern Strategy has rightly received enormous attention, it is often portrayed, via Kevin Phillips and Pat Buchanan, as a smashingly successful appeal to the white working class. According to historian Matthew Lassiter, this understates the extent to which the Southern Strategy was aimed squarely at the suburbs and was successful only insofar as it increased the share of the Southern white vote in presidential elections (Lassiter 2006; see also Davis 1986).

In short, the dog-whistles of the Southern Strategy were quickly recognized as such, and a complementary language that would appeal to Americans anxious about the upheaval of the 1960s but wary of the vestiges of segregationism (and their impacts on local economies) became increasingly vital to conservative causes. The "forgotten Americans"—a discursive appeal with a long history in US politics, drawn into the New Right via Goldwater—soon found its way into Nixon's public pronouncements. In his 1968 speech accepting the Republican Party nomination for president, Nixon spoke of "another voice" that was scarcely heard beneath the shouting of the moment: "It is the voice of the great majority of Americans, the forgotten Americans, the non-shouters, the non-demonstrators" (see Lowndes 2008, 113). The Nixon administration would soon pivot to similar symbols that were shorthand for the same demographic: the "silent majority" who inhabited the "middle America" so frequently ignored or rebuffed by cultural elites and New Left protestors alike:

> Richard Nixon called suburban families the Forgotten Americans, and then the Silent Majority, and finally the New American Majority. As populist appeals to Middle America, these labels represented a suburban strategy designed to conceal class divisions among white voters while taking advantage of the convergence of southern and national politics. (Lassiter 2006, 5)

Each of these ideal-typical concepts focused on a vision of normality; each was contrasted in opposition to a cultural 'Other' (the coastal elites, the loud minority, etc.); and each was colorblind on the surface but was clearly invoked and intended to appeal to white America (Rieder 1989, 244; Kazin 1998; Olson 2008). Agnew mused, in one speech, that "the forgotten American . . . does not enjoy being called a bigot for wanting his children to go to a public school in their own neighborhood" (Coyne 1972, 381; as cited by Olson 2008, 711). As wunderkind Republican strategist Kevin Phillips described in a 1970 editorial in the *Washington Post*, "The fulcrum of Republican appeal is more or less the 'social issue'—law and order, permissiveness, campus anarchy, racial engineering." Still, given the political dynamics of the time, Nixon needed to hedge his bets; so law and order came to rest uneasily alongside commitments to "liberal" causes. The Silent Majority may have been wary of forced busing and angry at antiwar protesters, but, for many Americans, the ecological troubles of an "affluent society" had also hit close to home.

Defenders of the Middle-Class Environment: The New Environmentalists

Beginning in the mid-1950s and intensifying over the course of the ensuing decade and a half, a strong current of thought declared that the suburbs were fast becoming ecological wastelands. In a biting critique of suburbia, John Keats referred to suburbs as "fresh air slums." He mused that the "typical post-war development operator . . . whistled up the bulldozers to knock down all the trees, bat the lumps off the terrain, and level the ensuing desolation" (1956, 5). In a series of 1954 *Harpers* articles, Frederick Allen made a case that, in the suburbs, access to nature was being democratized, but the unanticipated end result was ecological destruction. In the aptly named *God's Own Junkyard*, Blake (1964) argued that "the suburbs had lost the very qualities that people left the cities to find—privacy and free and open out-door space." Perhaps most ominously, a *Saturday Evening Post* article, brashly titled "The Rape of the Land," pointed out that "as more and more Americans rush to the few remaining convenient open spaces to build their homes or to take vacations, they destroy what they seek" (Bagdikian 1966).

On one hand, these portrayals of suburbia as an ecological wasteland were overstated, part and parcel of the liberal elite's disdain for the mundane culture and lifestyles being adopted by the middle-class masses. "Suburbia," wrote Bennett Berger, "has become a mass phenomenon and hence prone to the manufacture of modern myth" (1960, 1). The myth of the suburban environmental wasteland, Christopher Sellers contends, is "inattentive to the class and racial privilege in which it has so often been grounded, and . . . dismissive of places and experiences that less well-off suburbanites have considered natural" (2012, 41). On the other hand, though, the environmental problems of suburbia were not altogether illusory: nature *was* being transformed, with deleterious consequences in many cases. As historian Adam Rome has detailed, as suburban sprawl proceeded, the construction of new developments increasingly occurred in ecologically sensitive areas, like "wetlands, steep hillsides and floodplains" (2001, 3). Detergents and improperly installed septic tanks leached into water supplies and soils. The spraying of pesticides caused bird die-offs and provoked public health panics. Water supplies had to be secured, often from central cities or more distant locales. And consumer waste had to be disposed of—usually somewhere else (Rome 2001). For many observers, the suburbs reflected a dark side of the affluent society—the growth of mass consumerism and crass materialism.

Among critics detailing the environmental destruction of suburbia, the impacts of rapidly expanding automobility received particular attention. For suburb dwellers, wrote Wood, "the private automobile and the truck loom as the major channels of access to the outside world" (1958, 246). In *The Quiet Crisis*, Stewart Udall mused that "this flight to the suburbs—in part a protest against the erosion of the urban milieu—has had its element of irony, for the exodus has intensified our reliance on the automobile and freeway as indispensable to modern life" (1963, 172–173). In *The Affluent Society*, J. K. Galbraith offered a depressing picture of this new suburban American dream:

> The family which takes its mauve and cerise, air-conditioned, power-steered and power-braked automobile out for a tour passes through cities that are badly paved, made hideous by litter, blighted buildings, billboards and posts for wires that should long since have been put underground. They pass on into a countryside that has been rendered largely invisible by commercial art. . . . They picnic on exquisitely packaged food from a portable icebox by a polluted stream and go on to spend the night at a park which is a menace to public health and morals. Just before dozing off on an air mattress, beneath a nylon tent, amid the stench of decaying refuse, they may reflect vaguely on the curious unevenness of their blessings. Is this, indeed, the American genius? (Galbraith 1998 [1958], 187–8)

In a single decade, from 1945 and 1955, "automobile registrations doubled from 26 million to 52 million" (Huber 2013, 73). Between 1940 and 1960, the percentage of American families owning cars rose from 51 percent to 77 percent (Wells 2012, 279). Suburbanites were more likely to have cars than the average American, and to drive those cars more than the average car-owner (Wells 2012, 279–286; see also Wood 1958, 247). In the suburbs, cars were needed not only to commute to work, but for trips to the grocery store, shopping mall, secondary schools, and access to recreational activities. As Wells concluded, "Cars made suburban development possible, but mushrooming suburban developments made cars essential" (2012, 286). The post–World War II highway boom, the growth of sprawl, and the vastly increased burning of fossil fuels for auto use—all subsidized by the state—produced a legion of environmental consequences that themselves galvanized a social backlash.

Local (and, eventually, national) forms of political organization emerged to confront suburban ecological decline. Andrew Hurley notes that the Gary, Indiana, chapter of the League of Women Voters, stocked with suburban residents of Glen Park, Miller, and Horace Mann, "tapped into the growing discontent among women trapped in a revolving door of shopping, cooking, cleaning, and chauffeuring children" in their campaigns for smoke abatement and "pure water" (1995, 57–59). Another widespread movement was focused around groundwater pollution from septic tanks. Of particular concern was alkyl benzene sulfonate (ABS), a chemical used in detergents that was seeping from septic tanks into private wells. Rome finds that, in the early 1960s, "in magazines devoted to the concerns of middle-class women, from *Redbook* to *Good Housekeeping* to *American Home*, the subject was especially prominent" (Rome 2001, 107; see also Sellers 2012, 119). A 1963 *American Home* article connected this problem to the broader increase in water pollution: "The disturbing discoverings about the outbreaks of infectious hepatitis and the distasteful detergent suds problems plaguing many communities should drive home to all of us . . . the perils to family health from polluted water are steadily increasing" (Senn 1963, 45). From detergent-contaminated drinking water in Long Island to sewage-polluted wells in an upscale Connecticut suburb to waste-flooded lawns from septic tanks in a St. Louis suburb, the article linked looming water pollution to the decimation of suburban life. Such stories caught the attention of national media and policymakers, helping to galvanize support for federal water pollution standards in the process.

Alongside water pollution, open space emerged as a major suburban environmental concern. Rome observes that "in many communities, the advance of development led to the formation of grassroots groups, often led by women, to 'save our trees' or 'stop the rape of the valley'" (2001, 123). Centering their arguments around open space as a consumer amenity and a site of recreation, these groups took aim at overdevelopment and sprawl. In the *Saturday Evening Post*, journalist Ben Bagdikian remarked that "in every city and in thousands of towns and obscure neighborhoods, there are housewives and homeowners banding together to fight, block by block, sometimes tree by tree, to save a small hill, a tiny brook, a stand of maples." As one activist put it, "Any unspoiled natural area in any suburb is threatened today and will be spoiled tomorrow, unless someone starts fighting for it right now" (Paine 1958). Efforts to protect clean water and preserve open space were increasingly linked with the dangers of pesticides and other synthetic chemicals. Most notably, suburban activists in Nassau and Suffolk Counties, on Long Island, waged a battle against the spraying of DDT to eliminate mosquitos and gypsy moths. This activism resulted in well-publicized trials against DDT usage in 1957 and 1966, influenced Rachel Carson's best-selling *Silent Spring*, and eventually led to the formation of the Environmental Defense Fund (Sellers 2012, 133–136, 270).

The resistance that emerged out of local concerns over social and ecological reproduction—related to human health, and the nonhuman species and systems present in backyards and neighborhoods—was gradually collected under the banner of an emergent movement: environmentalism. The conceptual transformation from the *nature* of the old conservationism—overwhelmingly comprising elites and technocratic experts, focused on the scientific management of water, forests, soils, and wildlife—to the *environment* of the new movement was profoundly consequential: "A bodily reconfiguration of conservation's agenda brought its core concerns closer to home, made saving nature potentially relevant and applicable to far more suburban dwellers" (Sellers 2012, 133; see also Bond 2022).[14] "Environmentalism," observes Kirkpatrick Sale, "moved from being the concern of the affluent and elderly of the boardroom on the one hand or of the backwoods hunters and fishers on the other, to being the stuff of everyday life—and politics—for millions" (Sale 1993, 14). According to Sellers, "The leaders of these self-styled 'environmental' groups were often non-working housewives living in wealthy suburbs" (2012, 270). Bagdikian's aforementioned report included the story of children and housewives in Fairfax, Virginia, who had formed the Holmes Run Recreation Association in order to prevent 276 acres of woods from being transformed into a new development. One woman, a writer of children's books, took her case to Lady Bird Johnson, and received the following reply: "If I were to change the names and change the figures in your message, it would correspond with truly hundreds of letters that have come to me from suburban counties all over the nation."[15]

Suburban environmental struggles were, of course, by no means innocuous and sometimes outright exclusionary in both composition and ideology. The 1957 lawsuit against DDT use on Long Island was organized by an overwhelmingly elite cast of north shore residents who expressed little concern for the environmental despoilation occurring in working-class suburban locales (Sellers 2012, 99–103, 128–133). As the environmental movement exploded in the late 1960s and early 1970s, efforts to defend the suburban environment from ecological and cultural decline often overlapped—to such an extent that it was, in some cases, difficult to identify which impulse was driving the organizations seeking to prevent further suburban growth (Ward 2019). As a state legislator from the Atlanta area put it, "The suburbanite says to himself, 'The reason I worked for so many years was to get away from pollution, bad schools and crime, and I'll be damned if I'll see it all follow me'" (Kruse 2005, 247). In her analysis of liberal political activism in suburban Boston, Lily Geismer observes that "the localist measures that residents took to protect their communities elevated both a sense of their own distinctiveness and a focus on their own individual standard of living and quality of life, further obscuring an acknowledgment of their role in perpetuating many of the problems of environmental and social inequity" (2015, 98). Hurley similarly notes that in the suburban neighborhood of Miller, Indiana, where few residents were reliant on industrial jobs, a powerful homeowner's group eventually came to oppose "any sort of development that threatened to alter the neighborhood's character—trailer parks, apartment complexes, high density recreation facilities . . . often citing environmental disruption as a justification for their position" (1995, 75).

In spite of these shortcomings, the suburbanites active in the environmental struggles of the 1960s and early 1970s did more to branch out than their conservationist forebearers of the 1950s, at times participating in coalitions to protect the environment in urban and rural locations alike. Sellers details how umbrella groups, like Action for a Better Los Angeles (ABLE) and the Long Island Environmental Council (LIEC), concerned themselves with topics that spilled beyond the comfortable confines of upper-middle-class suburbia—including noise pollution and public transportation—and broadened their constituencies to include the working class and people of color (Sellers 2012, 270–271).[16] These more progressive organizational linkages were undercut, however, by the symbolic role that the suburban environment played as an ideological archetype—the normative standard for US social life—that was quickly adopted by both an ostensibly liberal social bloc that would populate environmental organizations in the years to come and a reactionary conservatism that would come to dominate the former party of Lincoln. While the treadmill of production expanded, the accompanying acceleration of the treadmill of social reproduction—a terrain constituted by racial and class-based segregation and a gendered division of labor—engendered cries of resistance that emanated loudly from suburbia. As the suburban voting bloc grew in size and strength, the demographic and political variability of the actually existing suburbs was drowned out by new political signifiers that functioned to transform suburbia into a synecdoche—the social and ecological setting in which the prototypical median voter resided—the environment in which American values were nurtured.

THE TWO TREADMILLS OF THE MIDDLE-CLASS ECOLOGICAL ORDER

The suburban environment was the residential foundation for the middle-class ecological order of the mid-twentieth century; a collection of real spaces that further entrenched, in different ways, racial, class, and gendered divisions in US life, and a spatial imaginary that erased the class struggles going on in the industrial factories, fields, and mines (and, to a lesser extent, schools, the transportation sector, and white-collar workplaces). From the preceding analysis, we can identify four defining characteristics of the mid-twentieth-century middle-class ecological order.

First, the middle-class ecological order emerged at a moment where a gulf had opened up between dominant concepts of class and actually existing class relations. On the one hand, the theoretical apparatus of classical Marxism *did* seem ill-equipped to make sense of the changing jobs and material conditions of many US workers who occupied "white-collar" or supervisory jobs or who were the blue-collar beneficiaries of the victories of labor (pensions, paid vacations, annual cost-of-living adjustments, etc.). Blue-collar jobs in mining, transportation, and heavy industry were declining, while positions in service, retail, and municipal and federal government were steadily increasing, and new occupations in the FIRE (finance, insurance, and real estate) sector were already rapidly growing (Dray 2011, 514; see also Boggs 1963). On the other hand, though, the rearguard strikes by capital and the state suggested that a working class-for-itself loomed larger in this period than at perhaps any previous point in US history. The hard-fought gains of the US labor movement had provoked an unparalleled attack on the working class by corporate America and its political extensions—like the National Association of Manufacturers, the Chamber of Commerce, the Manufacturing Chemists' Association, the American Petroleum Institute, and the Automobile Manufacturers Association. With automation lurking just around the bend, the 1947 Taft-Hartley Act dulled the toolbox of labor—outlawing the closed shop, making wildcat and solidarity strikes illegal, banning supervisory unionism, requiring union officers to sign anticommunist affidavits, and authorizing Right to Work legislation that incentivized free-riders and made organizing campaigns more difficult to wage (Davis 1986; Lichtenstein 1997; Dray 2011).[17] As ToP scholars pointed out, workers were chastened not only by the always-present machinations of capital accumulation but by the resounding victories of capital and the resulting constraints placed on organized labor by the state. This was compounded by federal policy, like the Housing Act of 1949, which—thanks to the success that the corporate lobby had in influencing Truman—was reduced to "a subsidy for business and middle-class homeowners, rather than the public housing program for the working-class which the CIO had originally envisaged" (Davis 1986, 94).

Labor writ large suffered from these attacks,[18] but the burdens fell disproportionately on working-class people of color who, generally speaking, had less seniority, were underrepresented in union leadership and the ranks of skilled labor, and toiled in "the meanest and dirtiest jobs" (Sugrue 1996; Marable 2007; see also Boggs 1963; Rector 2022).[19] In the agricultural fields, migrant farmworkers received meager wages while struggling against the growers, as well as enduring exposure to DDT and other pesticides (Pulido 1996, 79). Navajo, Ute, and Laguna Pueblo miners faced not only the health and safety hazards that plagued all miners, but the acute long-term toxicity of the uranium ore that they extracted and the open pit mines, tailings piles, and radioactive dust that they lived alongside on tribal territory (LaDuke 1999, 97; see also Voyles 2015).[20] In Michigan's auto factories, the workforce of foundries disproportionately comprised Black workers, who were daily subject to burns, extreme heat, and "the absorption of toxic metals" (Sugrue 1996; Stein 1998; Rector 2022). As a UAW industrial hygienist noted, in 1961, "In view of all of these conditions . . . it seems quite inevitable that foundrymen will normally have substantially shorter life expectancy than will most members of the working community" (Van Atta, as cited by Rector 2022, 96; see also Boggs 1963; Stein 1998; Ross and Amter 2010).

Deindustrialization in urban centers, combined with residential segregation in suburbia, added fuel to the fire of these racial inequalities. Workers of color were shut out of not only homes but also job opportunities in suburbia. For example, a 1954 *Time* magazine article noted that, "of 2,658 plants built in the New York area from 1946 to 1951, only 593 went up in the city proper." Historian Josiah Rector reports that, over the course of the 1950s, "840 factories closed in Detroit, while 124 new factories opened in the suburbs" (2017, 117). Geismer found that "a string of major textile factory closures" in the cities of Massachusetts after World War II was accompanied by a bonanza of tech-industry office parks in suburbia (fueled in large part by Department of Defense contracts) (2015, 21–23). "With few exceptions," wrote Geismer, "the office parks . . . lacked access to adequate public transportation and likewise most transit schedules served suburban commuters going into the city instead of the reverse" (2015, 23).

Second, in contrast with the mid-nineteenth-century class ecological order, the forms of racial discrimination that characterized the middle-class ecological order were not legitimated *primarily* by appeals to a "natural" racial order.[21] In the civil rights era, forms of biological racism were increasingly seen as social and political liabilities; instead, racial hierarchies were reinscribed in the environment. As eugenics was cast aside and the civil rights movement marched forward, racial differences that were once viewed as innate were now seen as deriving from a particular material setting—the socio-ecological relations that characterized one's day-to-day surroundings. The infamous 1965 Moynihan Report, for example, explicitly rejected biological racism: "There is absolutely no question of any genetic differential: Intelligence potential is distributed among Negro infants in the same proportion as among Icelanders or Chinese or any other group. American society, however, impairs the Negro potential" (US Department of Labor 1965, 35). According to this influential line of thinking, derived from earlier frameworks for understanding ethnic minorities (Taylor 2016, 36–42; Reed 2020, 98–112),[22] the historical realities of US racism had seeped all the way into the structures of "lower-class" African American families, resulting in a matriarchal culture that, as a consequence of its subordinate position within the dominant cultural fabric of the United States, created the conditions for pathological behavior, like juvenile delinquency, to flourish. Deepening residential segregation caused this "tangle of pathologies" that was said to be characteristic of poor and working-class African American families and threatened to spread like a virus:

> The children of middle class Negroes often as not must grow up in, or next to the slums, an experience almost unknown to white middle class children. They are therefore constantly exposed to the pathology of the disturbed group and constantly in danger of being drawn into it. (US Department of Labor 1965, 29–30)

In addition to the great irony that public intellectuals and critics of the time also routinely bemoaned the matriarchal structure of white middle-class suburbia (see, e.g., Fromm 2002 [1955], 155–156), it bears noting that the supposedly matriarchal culture of African American families was only one difference that characterized popular depictions of US races. In contrast with the suburban environment—quiet tree-lined streets with large yards, clean air, and a homogeneous architecture of cape cods and raised ranches—the "ghetto" or "urban slum" environment was typified by crime, noise, the absence of nature, and the presence of a deteriorating built environment. For example, a 1970 article in *Time* magazine, titled, "Ecology of a Ghetto," described the "ecological pattern" that was emerging as "blacks expand into Chicago's white areas":

> White workers leave; industries are not satisfied with unskilled blacks and increasingly move to the suburbs. Since blacks have less to spend, many stores also move. . . . As local jobs decline, Chicago's blacks must spend more—not only for higher rents, but also to travel farther for work. Only about 15% have cars. In addition, the most available jobs for ghetto blacks are the city's worst: janitor, forklift loader, punch-press operator, hospital orderly. . . . Consider Chicago's Near West Side. . . . Closer inspection reveals a streetscape of despair: low, glum buildings, boarded up store fronts, infrequent parks, broken curbs.

At the end of the article, one resident of inner-city Chicago angrily noted, "We don't have a new tree here. We don't have grass."

Like the suburbs, the ghetto was a place defined by its environment, and the resident's aforementioned description of the environmental problems plaguing the inner city wasn't off-base. The problem was that, among intellectuals and the media alike, "the environment" was being translated into a natural outgrowth of vast cultural differences. The dominant line of thinking held that from these distinct environmental milieus arose cultural norms and values that could then be attributed to particular races and classes. The suburban environment was the domain of the white middle class; as one 1959 Department of Labor study put it, the husband was "the architect of family fiscal policy" and the housewife was the "purchasing agent" (cited by Cohen 2004, 148). The ghetto environment was the stomping grounds of the Black underclass; the husband was depicted as absent and the wife or single mother, slightly more educated and possessing better employment prospects, was the authority figure (US Department of Labor 1965, 30–36). The Kerner Commission report—progressive in many respects—also made explicit these linkages between environment, culture, and race, arguing that the "culture of poverty" and resulting fracturing of the family that characterized the ghetto "generates a system of ruthless exploitative relationships. . . . Prostitution, dope addiction, and crime create an environmental 'jungle' characterized by personal insecurity and tension" (1968, 7). "Children growing up in such conditions," the report continued, "are likely participants in civil disorder." Perhaps this was so, but it had little to do with culture and everything to do with the dire material conditions of the inner cities, and the predictable desire that the inhabitants of such a place had for real change.

The environmental movement of the late 1960s and early 1970s echoed this thinking most explicitly through its embrace of neo-Malthusianism as a prism through which to understand the "urban crisis." Paul and Anne Ehrlich's best-selling *Population Bomb* (1968) contained a fictional scenario in which resource shortages caused by overpopulation lead to racial strife in the ghetto:

> Margaret Andrews had had very few choices in her life since Richard had been killed in the riots. He had died because of the things she had loved him for; his refusal to knuckle under to the dominant white society and, especially, his feeling of community with the oppressed people of the Third World. . . . The clarity with which the Population Control Law was aimed at the blacks and the poor had been the last straw. Even though they had carefully planned their two children, Richard had refused to speak out against the cries of revolution in the ghetto high school where he taught history. His patience was at an end, and his life soon ended also, snuffed out by a random bullet fired in the worst civil disorder in the history of the United States. (54)

More often than not, though, environmentalism reinforced this ideological divide between the suburban and ghetto environment in a more subtle and simple fashion: through its near-exclusive focus on the suburban citizen. With some notable exceptions, environmentalists overwhelmingly wrote, organized, and acted with the middle-class suburban environmental subject in mind, and lavished attention on those spaces assumed to be the most natural: the wilderness, then the upper-middle-class suburbs, with the industrial suburbs and cities an afterthought or a case study in what can go wrong.

This tended to result in an anemic view of both what constituted an environmental issue and where there was nature worth preserving. When suburban environmentalists were fighting against pesticide exposure in the mid-1960s, for example, the United Farmworkers were waging their own struggle against the impacts of DDT sprayed in the fields (Pulido 1996). While the UAW and International Longshoremen's and Warehousemen's Union stood in solidarity with the striking farmworkers, it was much more difficult to convince middle-class environmentalists that "defending the natural world and human health demanded engagement with campaigns for social justice" (Montrie 2011, 125–126).[23] Along similar lines, compromises taken by the Sierra Club to protect the crown jewels of American wilderness at times produced profound environmental injustices. The club long supported the mining and burning of coal as an alternative to the construction of big dams providing hydropower; in the Southwest, this directly displaced environmental destruction away from the wild Colorado River (so that its flows through the Grand Canyon would be preserved), and onto the Navajo reservation, where coal would be mined and burned to provide power to the rapidly expanding cities and suburbs of the Southwest (see Needham 2014, chap. 6).[24] As Sellers argues, "Environmentalists' search for salves to their own alienation, for closeness to a nature that seemed purer and more genuine to them, kept the altruistic potential of their new movement mostly unborn" (Sellers 2012, 247; see also Ward 2019).

The natural wages of whiteness of the mid-nineteenth century had morphed into environmental wages of whiteness by the mid-twentieth: the residential mobility afforded to middle-class whites enabled them to insulate themselves from some of the most egregious and intolerable environmental harms by relocating to the suburbs, and, in the event that this harm followed them, their relative access to political resources enabled them to mobilize in defense of the suburban environment;[25] the distinction between the suburban and ghetto environment naturalized cultural difference in racial terms, elevating the suburban environment to the normative standard while linking the disproportionately poor and racial minority residents of urban centers with delinquency, crime, and deviance; and this residential segregation was widely taken to be a product not of discriminatory policy and practices, but of the free choices—for some commentators, even natural tendencies—of individuals representing distinct cultural groupings.

Third, much like the wild frontier of the mid-nineteenth century, the middle-class ecological order of the mid-twentieth was dominated by an ideological construct—the suburban environment—that veiled material social and ecological relations. The real suburban environment was not home to a burgeoning bourgeoisie, but a "contradictory location within class relations." This concept, coined by Erik Olin Wright (1980), refers to the fact that many "middle-class" workers occupied ambiguous positions within relations of production. While certainly not owners of the means of production, supervisory and managerial workers were often delegated some measure of authority over the production process (and the workers on the shop floor), relative autonomy in charting out their day-to-day schedules, and wages and benefits that enabled a lifestyle of middle-class consumerism. For such a worker, who has some interests that align with capital and others that align with labor, political intervention and ideology becomes increasingly important in adjudicating which side of political struggles they fall on. The suburbs were home to many such workers, and their politics were subject to the scrutiny of scholars, the media, and both major political parties.[26]

Although labor unions continued to wage consequential battles, the residents of suburbia increasingly viewed their interests as more than the sum of their paychecks, and their politics as distinct from their vocation. In his history of American unions, Phillip Dray argues that, by the 1950s and 1960s, "with workers commuting to their jobs each day by car, a workforce no longer needed to reside within the shadow of the mill or plant, nor did men congregate in union halls or gather over beers each night to swap workplace grievances; increasingly they perceived their lives and their aspirations as things separate from their means of earning a paycheck" (Dray 2011, 515). The suburbs became core sites of political struggle on the ground, and an essential target demographic for both major political parties. While the suburbs were varied terrain—demographically, geographically, and politically—the suburban environment acquired an ideological content that was sanitized of this on-the-ground messiness. This is most notable in the Republican Party's efforts to construct a discursive bridge between the suburbs and the Silent Majority, but it would also become integral to the Democratic Party's hard right turn in the latter half of the 1970s and early 1980s (described in more detail in chapter 3). Unfortunately, the political Left too frequently wrote off the suburbs as politically predetermined by the apparently bourgeois realities of private homeownership and consumer culture. This self-defeating tendency led Bennett Berger to muse that "it is almost as if left-wing social critics feared the seduction of the working-class by pie not in the sky, not even on the table, but in the freezer" (Berger 1960, 103).

Fourth, the material and ideological content of the suburban environment did nonetheless impose real barriers to political organizing, making it difficult to forge sustained linkages between middle-class environmentalists and the more radical political movements of the era. Not surprisingly, labor unions had a complicated relationship to environmentalism: the AFL-CIO supported the Wilderness Act, the UAW contributed funding for the first Earth Day, and a variety of unions, like the Oil, Chemical and Atomic Workers (OCAW), Steel Workers, and United Farmworkers, began to prioritize environmental health and safety in the 1960s and early 1970s (Dewey 1998; Gordon 1998; Obach 2004, chap. 3; Montrie 2011; Rector 2014, 2017).

Environmentalists, in turn, occasionally joined in solidarity with labor. In 1973, for example, when the OCAW struck over workplace health and safety, eleven major environmental organizations stood in solidarity with the oil workers (Obach 2004, 50). A few years later, environmentalists joined unions in beating back a bill to gut OSHA (Kazis and Grossman 1982, 90). At the same time, however, organized environmentalism emerged primarily on the terrain of leisure rather than labor; it was given shape by the call of the wild and the shout of the suburban household, not the chants of striking workers. As capital began to proclaim its fervent opposition to all things environmental, emphasizing potential plant closures and job losses, unions grew understandably concerned. The politics of the treadmill complicated—and, in many cases, decimated—blue-green coalitions, adding to the hardships faced by the working class and the social reproductive burdens disproportionately born by working and middle-class women. The resulting environmental mobilizations emerged from the home and neighborhood rather than the factory floor. The struggle to safeguard the environmental wages of whiteness, while certainly evident in ongoing workplace discrimination against women and racial minorities, was most glaring in the homes and neighborhoods of suburbia, where movements—conservative and liberal alike—fought to preserve their natural and cultural environment by insulating themselves from external forces and perceived threats.

As a consequence, while some African Americans did join the first Earth Day and became active in environmental causes, cooperation between civil rights activists and environmentalists was largely fleeting, and the interactions between the two movements occasionally hostile. As Robert Bullard has argued, "Poor and minority residents saw environmentalism as a disguise for oppression and as another 'elitist' movement." (1990, 9, 26). Although many activists attempted to build bridges between environmentalism, labor, and civil rights, the sentiment was widespread.[27] Nathan Hare mused that "the ecology crisis arose when the white bourgeoisie, who have seemed to regard the presence of blacks as a kind of pollution, discovered that a sample of what they and their rulers had done to the ghetto would follow them to the suburb" (1971, 4–8). During Earth Day 1970 at San Jose State, as young white environmentalists exuberantly buried a car, Black students protested in shock against the wasteful and privileged act (see, e.g., Hurley 1995, 10). The overwhelmingly white middle-class composition of environmentalism led Richard Hatcher, the African American mayor of Gary, Indiana, to muse that "the nation's concern with environment has done what George Wallace was unable to do: distract the nation from the human problems of black and brown Americans." Hatcher's critique of environmentalism, nonetheless, was nuanced: "The mayor reasoned that a nation wealthy enough to send spaceships to the moon surely had the resources to combat both poverty and pollution" (Hurley 1995, 111). And, "when prominent African American leaders from across the nation convened in Gary in 1972 to chart a course for independent black politics, they included several planks about industrial pollution in their manifesto for change" (Hurley 1995, 12). The opposition of most African American activists was not to environmental protection, but to the failure of the environmental movement to adequately integrate what we now know as environmental justice concerns into its activism (Furgurson 1970; Hare 1971; Tingling 1980).

The most vehement resistance to environmentalism, by far, came not from unions or civil rights activists, but from capital. Already, by the mid-1940s, the LA Chamber of Commerce, joined by auto manufacturers, the railroad and lumber industries, and the Western Oil and Gas Association, had objected to early local efforts to control air pollution and lessen smog (Ross and Amter 2010, 76–81; Esposito 1970, chap. 2). As water and air pollution standards gradually worked their way into state and national legislation over the course of the 1950s and 1960s, this opposition—though at times tempered by the grim realities of pollution, and the weakness of the proposed laws—failed to abate in any meaningful way: industry struggled against local, state, and national antipollution laws; fought to delay and weaken the laws if they were passed; touted the money spent on pollution-control technology while doing little research that produced tangible results; and, of course, threatened workers and communities with massive layoffs and even plant closures if effective environmental laws were institutionalized.[28] In a 1970 article titled, "The Rise of Anti-Ecology," *Time* magazine reported,

> U.S. Steel Corp. has threatened to close all its plants in Duluth rather than spend $8,000,000 for pollution controls required by the state. A shutdown, city fathers fear, would throw 2,500 people out of work and severely damage the city's economy. . . . B.A.S.F., an American subsidiary of a large German chemical company, has suspended plans to build a $200 million plastics and dye complex in poverty-stricken Beaufort County, S.C., until it determines just how expensive Government-ordered pollution controls will be.

In *Vanishing Air*, published by Ralph Nader's study group on air pollution, John Esposito noted that General Motors "claim[s] to be spending about forty million dollars annually on research related to air pollution control . . . [but] the thrust of the company's effort is toward discouraging talk of alternatives to the internal combustion engine, rather than searching earnestly for new propulsion sources" (1970, 28). Esposito observed that,

> in 1967, virtually every industry witness who testified concerned the pending Air Quality Act urged Congress to drop an administration proposal which . . . would have given the Secretary of Health, Education and Welfare (HEW) authority to require that polluters disclose what they were putting into the atmosphere. . . . Among these associations were the Manufacturing Chemists Association, American Paper Institute, American Petroleum Institute, National Coal Association, American Mining Congress, Edison Electric Institute, the

U.S. Chamber of Commerce, and the National Association of Manufacturers. As a consequence, the Air Quality Act of 1967 contained no provision for compulsory disclosure. (1970, 75)

The arguments of the industrial lobby were repeated verbatim by the ideological mouthpieces of US conservatism. Bemoaning the high emotions and higher spending going into pollution abatement, a 1968 article in the *Public Interest* declared that cleanliness is relative, and some bodies of water, like Lake Erie or the Potomac River, may never be suitable for swimming, fishing, or drinking (Starr and Carlson 1968, 131).[29] In the *National Review*, Robert Moses called the ecology movement a new form of "public hysteria," declaring that "we have been brainwashed by sensational stuff and militant minorities" (1970). The Daughters of the American Revolution passed a resolution declaring the first Earth Day—which "not coincidentally" fell on Lenin's birthday—to be "subversive." "Pollution," they proclaimed, was "being distorted and exaggerated by emotional declarations and by intensive propaganda'" (*Time* 1970).

From the perspective of the treadmill of production, it's clear that such arguments obscured what was actually happening: while consumer demand had increased dramatically, capital was appropriating massive swathes of nature—at home and abroad—for chemically and energy-intensive production practices, and then releasing the voluminous by-products back into nature in the form of industrial pollution. The profits were channeled upward, while the risks rained down on the masses, provoking resistance in the process.

CONCLUSION

This chapter has attempted to provide insight into the class ecological order of the mid-twentieth century by exploring the prototypical environment of an emergent middle class—that of the suburbs. My analysis of the suburbs has hinged on two core concepts and their interrelations: the treadmill of production and what I've termed the treadmill of social reproduction. I've argued that the treadmill of production, in which increasing rates of surplus value were extracted from labor and reinvested in new forms of chemically and energy-intensive production, created widespread ecological devastation and intensified economic insecurity. These problems were difficult to evade, even in the relative comfort of suburbia, where their impacts were disproportionately borne by middle-class women engaged in the labor of social reproduction. The political mobilizations that emerged as the treadmill of social reproduction accelerated were characterized by a "defensive localism" (Sugrue 1996) that wove its way into both Republican and Democratic Party politics.

As the aforementioned White House Conference on Natural Beauty took place in 1965, the country was in the midst of a partisan realignment that witnessed conservative Southern Democrats jumping ship to the Republican Party as the gains of the civil rights movement were institutionalized in public policy. FitzGerald Bemiss, chair of the panel, "The New Suburbia," was among those Southern Democrats. In 1964, he had refused to endorse Johnson for president; by 1972, he had changed parties and was spearheading the Virginia Committee to Re-Elect President Nixon.[30] He would eventually leave the state Senate to preside over one of his family's businesses, the Virginia Sky-Line Company (which operated concessions along the scenic Sky Line Drive), and to serve on the board of directors of the James River Paper Company. His desire to conserve the state's natural resources was genuine, but his brand of environmentalism was one that existed alongside a robust commitment to capitalism and the "free market."[31]

His experience of suburbia was also far from that of the Levittown masses, or even the upper-middle-class women who mobilized against the use of DDT on Long Island. When Bemiss moved from Richmond's Monument Avenue to the outskirts of the city, he and his family resided in a stately mansion—described in one magazine article as "a grand house on a high bluff above the James River . . . that bears an amazing likeness to the Executive Mansion in Capitol Square" (Slipek 1980). Complete with minimum-acreage requirements, the exclusive Windsor Farms neighborhood in which he lived fit the image of the aristocratic suburb of yore—a village in the countryside originally "marketed to the well-to-do of Richmond to draw them out of the grime and crime of the city for bucolic country living without giving up the modern conveniences of fire, police, public sewer and water, access to doctors, transportation and schools of the city."[32]

While Bemiss sought to conserve a particular vision of the environment, his Windsor Farms neighbor, Lewis F. Powell Jr., thought it more appealing to wage a frontal attack on environmentalism. In August 1971, after a conversation with yet another Windsor Farms neighbor—US Chamber of Commerce executive Eugene B. Snydor—Powell wrote a memorandum to the chamber's Educational Committee, titled "Attack on American Free Enterprise System" (Robertson 2014). Detailing a laundry list of grievances against the New Left, campus activists, civil rights organizations, labor unions, and environmentalists, Powell mused that "business has shunted confrontation politics. . . . There should be no hesitation to attack the Naders, the Marcuses and others who openly seek destruction of the system." He continued, "Current examples of the impotency of business, and of the near-contempt with which businessmen's views are held, are the stampedes by politicians to support almost any legislation related to "consumerism" or to the "environment"" (Powell 1971).

To combat this impotence, Powell made the case for a massive corporate mobilization of resources: investments in conservative scholars and speakers, the publication of business-friendly books and advertisements, and—above all—unapologetic corporate lobbying. Powell's clarion call—reminiscent of Goldwater both in its faith in the market and its rejection of compromise—was heard loud and clear, and ultimately helped to lay the foundations for an anti-environmental movement that would blossom over the course of the 1970s. Only two months after writing the memo, which was not yet released to the public, President Nixon announced Powell's nomination to the Supreme Court.

The frayed nerves of well-heeled Windsor Farms residents were by no means characteristic of general middle-class suburban angst, but this example illustrates the contradictions of the suburban environment and highlights the connections between the national politics of the mid-twentieth century and the local suburban movements that I've discussed in this chapter. The twin movements of suburbia—the new conservatism and new conservationism—were, in spite of their myriad differences, linked by the disorienting class politics that produced them; they were, as Mike Davis so aptly put it, "class struggles of a third kind, involving neither militant labor nor reactionary capital, but insurgent middle strata" (1986, 228). As the example of Windsor Farms illuminates, however, this insurgent middle strata had no shortage of support from on high. The ideological ambiguity of the suburban environment was politically productive for the Right, enabling cross-class alliances between the Bemiss's of the world and the white- and blue-collar workers who had fled the social and ecological disarray of urban industrial centers. By contrast, an insular focus on a localized suburban environment too often constrained the vision and foreclosed the types of structural demands that could have enlivened the Left at a moment where labor was under attack and civil rights movements were struggling to translate legislative victories into real equality at home and at work.

The environmentalism of the era "did not rest on a monolithic social base" (Hurley 1995, 12), but the power of the treadmill of production made the trade-off between jobs and environmental protection a real one for the working class, particularly those who could not afford homes in the suburbs or who were excluded from the suburbs by the persistence of racial discrimination in lending and zoning practices (the latter of which were, by the 1970s, increasingly legitimated by ostensibly environmental concerns over growth). The suburban-inflected environmentalism that emerged during the movement's oft-invoked "golden age" would soon dovetail far too comfortably with the market-friendly politics of the New Democrats that I explore in chapter 3. To stretch the metaphor perhaps too far, we might say that rather than seeking to slow the treadmill of production or manufacture a treadmill of social reproduction that could be enjoyed by all, suburban political movements sought to stabilize the existing treadmill of social reproduction by periodically cleaning its belt and restricting its use to particular clientele. We continue to struggle against these tendencies today.

In September 2021, the Robert E. Lee monument on Monument Avenue in Richmond, Virginia, was finally torn down (Tavernise 2021); however, the defense mechanisms organized around the ideological monument to the bourgeoisie that I've examined in this chapter have remained frustratingly impervious to political pressure. The suburban environment continues to function as a rigid, yet resonant, trope—a shortcut for signifying an imagined American norm of middle-class comfort that politicians of all stripes invoke to conjure a memory of national flourishing. Contemporary political economic dislocations have encroached on and, in many cases, imposed great harm on the lives of actual people living in actual suburbs, but the monotonous stability of the suburban signifier—the "little boxes" on tree-lined streets with big yards and clean air—continues to allow titans of industry to craftily pose as mouthpieces for the suburban set, while reaping gross profits from the stabilization of particular natural places amid capitalism's continued assault on the broader environment.

That we're now witnessing the devastating impacts firsthand is not for lack of warning. Already, at the time of LBJ's environmental conference, there were signs of serious structural instability afoot. Three months prior to the White House Conference on Natural Beauty, Johnson had delivered a special message to Congress. Calling for a "new conservation" focused on "the total relation between man and the world around him," the message described the magnitude of the crisis and alluded to an ominous problem: "This generation has altered the composition of the atmosphere on a global scale through radioactive materials and a steady increase in carbon dioxide from the burning of fossil fuels."

3 "THOSE WHO BRING FROM THE EARTH": ANTI-ENVIRONMENTALISM AND THE CONSERVATIVE LEXICON OF CLASS

In the fall of 1980, against the backdrop of a presidential election, two very different political gatherings took place. From October 10 to 12, an organization called Environmentalists for Full Employment (EFFE) convened the First National Labor Conference for Safe Energy and Full Employment. Gathering at the Pittsburgh Hilton hotel, the nearly one thousand attendees, representing fifty-five different unions from thirty-three states, aimed to stake out an opposition to nuclear power and to consider the viability of alternative energy sources that could meet the needs of both workers and the environment. EFFE was formed in response to the exclusion of environmentalists from the national Full Employment Action Council, a coalition of civil rights leaders, labor unions, and women's rights activists, which sought to codify in law that there be "useful and rewarding employment opportunities for all adult Americans willing and able to work." The founders of EFFE believed that "without environmental safeguards," the Equal Opportunity and Full Employment (Humphrey-Hawkins) Act being pushed by the council "could become a license for untrammeled corporate production." In a letter to US Representative John Conyers, Richard Grossman, cofounder of EFFE, wrote, "We would like to see serious consideration given to energy conservation, labor-intensive industrialization, and to further protection of worker health and safety" (Grossman 1976).

The Safe Energy and Full Employment conference was nothing if not lively. At one point, one hundred workers from an International Brotherhood of Electrical Workers (IBEW) local stomped into the proceedings, carrying signs reading "no nukes are kooks" and "the nation needs nuclear power." As shouts rang out from the audience, the United Mining (UMWA) workers in attendance responded by marching up to the stage. As the conference seemed poised to devolve into prolonged in-fighting, the tension was broken as sounds of "Solidarity Forever" rang out, and the IBEW workers filed out of the rally (EFFE 1980). With participants united in opposition to the pro-nuclear contingent (with Three Mile Island in close proximity and fresh in everyone's minds), the subsequent discussions focused on laying the groundwork for understanding and advancing the mutually beneficial relationship between clean, safe energy and good jobs. As Rosemary Trump, international vice president of the Service Employees International Union (SEIU), mused, "What's good for Gulf and Westinghouse, it is becoming clear, is not good for us as workers. . . . But why can't we have safe energy and our jobs, too?" (Trump 1980). William Winpisinger, president of the International Association of Machinists and Aerospace Workers, echoed this sentiment:

> We feel we have a moral obligation to consider the public welfare on an issue involving life or death, health and safety. Our obligation doesn't end on the jobs site or with a pay check. . . . Let us turn our energies and imagination, the skill and knowledge of our members, to developing an energy agenda that is in harmony with full employment and economic democracy. (Winpisinger 1980)

The next month, nearly 2,000 miles due west, a very different discussion was taking place. From November 20 to 22, basking in the glow of Ronald Reagan's resounding victory over Jimmy Carter, roughly five hundred ranchers, farmers, and federal, state, and local officials convened at the Little America Hotel in Salt Lake City, Utah, to chart, as the conference was titled, "A New Federal Land Policy" for the 1980s. The conference, organized by the League for the Advancement of States' Equal Rights (LASER), was an attempt to capitalize on the momentum of what had come to be known as the Sagebrush Rebellion. Named after a shrub native to the US west, the movement arose on February 14, 1979, when AB 413 was introduced in the Nevada legislature, declaring state sovereignty over most federal lands in the state. Long outraged over perceived federal mismanagement of public lands, and aggravated by the impacts of recent environmental legislation on lands management policy, the "sagebrush rebels" sought—in blatant opposition to the US Constitution—to devolve control over public lands to state governments. "We're going to have a head-on confrontation. We're going to arrest all the BLM [Bureau of Land Management]," said Democratic state senator Norm Glaser (Rice 1979). While largely symbolic in terms of its actual legal impact, the Nevada bill passed, and the Sagebrush Rebellion quickly took hold across the US west, with similar measures enacted in Arizona, New Mexico, Wyoming and Utah, and a US Senate bill introduced that would have transferred 600 million acres of public lands to state land commissions (Thompson 2016).

Though it found ample support in many western communities, the rebellion was championed and financed by organized business and closely connected to politicians of the New Right. The Cattlemen's Association, the Farm Bureau Federation, and the oil and gas, coal, and forestry industries all voiced their support (Cawley 1993, 92–93; Helvarg 1994, 64–67).[1] Barry Goldwater served on the LASER advisory board alongside Senators Orrin Hatch (Utah) and Ted Stevens (Alaska), and the rebellion was vocally supported by the National Rifle Association. Attempting to capitalize on the popularity of such measures, Reagan had, during his presidential campaign, also joined the chorus of criticism aimed at federal land managers tasked with protecting and preserving public lands. On a June 1979 visit to Salt Lake City, he chimed in: "I happen to be one who cheers on and supports the Sagebrush Rebellion. . . . Count me in as a Rebel." Shortly after taking office, he sent a note to Sagebrush leader Dean Rhodes, which was read aloud to the cheering LASER conference attendees: "Please convey my best wishes to all my fellow sagebrush rebels" (*The Progress* 1980).

These two conferences laid out very different visions of the relationship between environmental protection and the well-being of US workers. The LASER conference provided the anti-environmental Right a bridge to connect the angry voices of the Sagebrush Rebellion to right-wing political and economic elites who desperately sought any semblance of connection to the common man. In this vein, the rebellion provided a grassroots springboard for a ruling class counterrevolution that, as I detailed in chapter 2, had been heating up over the course of the 1970s. The image of ruggedly individualistic, independent producers—white male cowboys and ranchers with deep roots in the frontier political imaginary—squaring off against federal bureaucrats furthered the construction of a broader national narrative that pitted ordinary, working-class Americans against a "new class" of environmental elitists. The fact that these ranchers were, by and large, not working class, and that many supporters of environmental laws and regulations *were*, was rendered invisible by the harsh political winds of the moment: the ascendance of financial capitalism, the deterioration of the New Deal coalition, and the mobilization of anti-environmental organizations (e.g., the Mountain States Legal Foundation, the Center for the Defense of Free Enterprise) and media sources (e.g., the *Washington Times* and the *Wall Street Journal*).

By contrast, Environmentalists for Full Employment was part of an attempt to forge a lasting coalition between unions and environmentalists—to cultivate a working-class environmentalism that reached across lines of race, nationality, and gender. Building on the momentum of the first conference, subsequent projects organized by EFFE were ambitious in the scope of participants and the substantive goals they sought to achieve. A March 1981 demonstration in Harrisburg, Pennsylvania, held on the second anniversary of the Three Mile Island (TMI) nuclear meltdown, brought together fifteen thousand people in efforts to keep TMI shut down, to win meaningful compensation for displaced workers, and "to demand jobs and job rights for all." The march was sponsored by twelve major unions and endorsed by the Sierra Club, Friends of the Earth, Mobilization for Survival, Ralph Nader's Critical Mass, Jesse Jackson's Operation PUSH, the National Organization for Women, the National Black United Front, and La Raza Unida. In his remarks to the crowd, Mike Olszanski, chairman of the Environmental Committee of the United Steel Workers, insisted that "the alternative sources of energy, sun and wind, are cheap, clean, and safe and would create lots of jobs. Nuclear power is unsafe and inefficient and eliminates jobs" (EFFE 1981). While setting the stage for further collaborations, the initiative ended in disappointment amid the political economic ruptures of the early 1980s.

By the end of Reagan's first term in office, Environmentalists for Full Employment had disbanded. The political vision of sagebrush activists, on the other hand, was being voiced in the halls of power. Though the rebels achieved only limited success in institutionalizing their policy goals, they provided an iconography of grievance and resentment that was quickly woven into conservative dogma: environmentalists were new class elitists, out of touch with the realities of ordinary Americans; extractive laborers, by contrast, were the vanguard of America's working class. This sentiment would eventually be codified in the national Republican Party Platform: "We are the party of America's growers, producers, farmers, ranchers, foresters, miners, commercial fishermen, and all those who bring from the earth the crops, minerals, energy, and the bounties of our seas" (Republican National Committee 2016). In previous chapters, I argued that the ideological tropes that worked to dull class consciousness—the independent producer of the wild frontier and the middle-class consumer of the suburban environment—were deployed by various political interests as more palatable alternatives to class politics; in this case, however, at a moment when liberals were turning their eyes away from class conflict, conservatives shifted their sights directly to the "American worker."

How did extractive workers climb to the pinnacle of US working-class imaginaries? This chapter attempts to answer that question by tracing the intersection of environmental politics and class struggle from the late 1970s to the early 2000s. I focus attention not on the epistemological counternarrative being woven by climate deniers—that is, their "scientific" claims, which have received ample attention elsewhere—but on (1) the political economic contours of the conjuncture in which organized anti-environmentalism sprang to the fore, (2) the relationship of anti-environmentalism to evolutions in both the Republican and Democratic Parties, and (3) the economic and cultural arguments that anti-environmentalists were articulating as climate change came crashing onto the public agenda. As the environmental justice movement was mobilizing to contest the unequal distribution of environmental harms and amenities, conservatives were crafting a political narrative in which environmentalists were tethered to "new class" elitists whose attempts at aiding the so-called underclass were part of a self-interested effort to insulate themselves from social upheavals by passing the burden off onto the working-class majority. In this narrative, the trope of the white male extractive worker became the master signifier around which a new conservative lexicon of class was being constructed. While this working-class imaginary was sited far from the suburbs that I discussed in chapter 2—in rural, western spaces where ranchers, farmers, roughnecks, and loggers subdued nature—it nonetheless resonated with sizable segments of a citizenry awash with anxieties about the direction of US politics and the future of their jobs and industries in a globalizing world.

THE CONTEXT FOR THE COUNTERREVOLUTION

To understand how such a strategy emerged, it is necessary to briefly survey the shifting political economic terrain of the 1970s. As Jefferson Cowie writes, "The years prior to the 1973–4 crisis had been the most economically egalitarian time in U.S. history" (2010, 12). For nearly forty years, wages had grown alongside productivity. Economist Judith Stein notes that, "after World War II, the economy grew 4 percent a year, and poor people gained more than the rich. The income of the lowest fifth increased 116 percent, while the top fifth grew 85 percent; the middle also gained more than the top" (2010, 1). For that reason, economists often refer to the period between World War II and the mid-1970s as the "great compression" (Goldin and Margo 1992). Working-class people were the prime beneficiaries of this compression: "Union driven spending funneled money from the company to the family and community. . . . The weekly earnings of non-supervisory workers increased 62% between 1947 and 1972 before stagnating indefinitely thereafter" (Cowie 2010, 28). As I detailed in chapter 2, this affluence was not distributed equally—far from it. In 1972, while the overall unemployment rate hovered around 5.5 percent, the African American unemployment rate was over 10 percent (Federal Reserve Bank of St. Louis 2023). In 1970, the average full-time female worker made roughly $30,000 (in today's dollars) while the average full-time male worker made nearly $50,000 (US Department of Labor 2018). The poverty rate of African Americans, Hispanics, and Native Americans remained between two and three times the white poverty rate (US Census Bureau 2020). These continued inequalities were profoundly unjust, but they had been diminishing over the previous decades, in some cases markedly. Broadly speaking, it was a time where the economy was growing, and the wealth created was being distributed more equitably than it had been before.

In this era of rising tides, the environmental movement was wildly popular. A 1968 article in the *Public Interest* commented that "if it is precise to describe the American people—that tangled collective—as having made a commitment to achieve any single public objective, it may be said fairly that they have resolved to clean up the waters of the nation and rid them of pollution" (Starr and Carlson 1968, 104). In the early 1980s, this continued to be the case: "There seems to be no issue, large or small, that can win more support, year in and year out, than the notion of 'protecting the environment'" (Tucker 1982, xiii). By this point in time, however, the political economic currents of the day had already begun cutting like a riptide through the placid waters on which such consensus was maintained, creating a path to the surface for corporate opposition that had long swirled just below. As Stein writes, "The 1970s was the only decade other than the 1930s wherein Americans ended up poorer than they began" (2010, 10). The path to this economic regression was paved by global political economic realities, both distal and proximal; the former including globalization, financialization, and deindustrialization, and the latter marked by the shift to floating exchange rates, oil shocks in 1973–1974 and 1979, and the phenomenon of "stagflation." These crises, emerging just as major environmental reforms were being institutionalized, immediately posed challenges to the realization of the ecological agenda, and ultimately served to strengthen the hand of capital in relation to both labor and environmentalists. As I expand on later in the chapter, the treadmill of production was being repackaged by Wall Street in a shiny new box that could be easily shipped across borders.

First, however, it is worth briefly discussing the direct sources of the backlash—two oil crises and the rise of inflation. By 1972, the United States produced 11.2 million barrels of oil a day but consumed 17.4 million (Stein 2010). In December of 1973, a barrel of crude oil cost $27; the next month it shot up to $63 (Macrotrends 2023). Front-page news stories told of long lines at the pumps, and working-class people struggled to keep up with the costs of gas and heating oil. Organized business took advantage of this opportunity to displace blame onto environmentalists, labor unions, and the federal bureaucracy. "We have legislated the energy crisis by a multiplicity of regulations and licensing requirements dealing with auto emissions, nuclear safety, mining operations, exploration for oil and gas, siting refineries, and many other energy related activities," read a 1973 article in the National Association of

Manufacturers journal (Meese 1973). Following this logic, the only solution to the energy crisis was to deregulate energy. While industry bemoaned Nixon's wage and price freezes, they applauded his termination of oil-import quotas and tariffs on imported oil, as well as his full-fledged support for construction of the Trans-Alaska Pipeline (National Association of Manufacturers 1973; see also Phillips-Fein 2009; Layzer 2012, 41).

The oil crisis helped to fuel inflation and led to a corresponding debt crisis in the developing world. From 1973 to 1974, the inflation rate nearly doubled in the United States (WorldData 2023). By 1980, amid a second oil crisis, rates of inflation again skyrocketed to nearly 15 percent. In curating policy responses to inflation, the corporate lobby, as well as many mainstream economists, unsurprisingly laid the blame at the feet of environmentalists and unions. Environmental and occupational health regulations were said to be "causing inflation, slowing growth, hampering inventions, hurting small business, making it difficult for American companies to complete on the international market and limiting capital formation" (Phillips-Fein 2009, 196). For example, amid the implementation of the 1977 Clean Air Act amendments, the Motor Vehicle Manufacturers, American Petroleum Institute, American Iron and Steel Institute, American Mining Congress, American Paper Institute, Construction Industry Advancement Fund, National Forest Products Association, Chevron, and the Chamber of Commerce held the misleadingly named National Air Quality Conference, where they argued that the law would "cause inflation, threaten jobs and stop economic growth" (Hornblower 1979).

Presidents Ford and Carter both took notice of the pleas from industry. Ford established the Council on Wage and Price Stability (which considered how environmental programs affected inflation) and required Inflationary Impact Statements for new rules and regulations (Kazis and Grossman 1982, 88; Layzer 2012, 43). Continuing the Nixon administration's attempts to quash the EPA from within the executive branch, such measures gave the Office of Management and Budget enormous leverage over EPA policy (see Kazis and Grossman 1982, 89; Layzer 2012, 39). Although Carter came into office touting the remarkable job growth that had accompanied environmental regulations, "as inflation worsened, [he] became more susceptible to the advice of his economic advisors and the business-oriented members of his cabinet . . .

who were pressuring the EPA and Interior Department to reduce the cost of regulation to fight inflation" (Layzer 2012, 69). Ultimately, he acceded to amendments to the Clean Air and Endangered Species Acts that significantly watered down both pieces of legislation, in part, in the name of lowering inflation and prices. He also issued Executive Order 12044, which required that all proposed regulations undergo extensive cost-benefit analysis; he relaxed regulations on logging in federal forests; and he allowed the EPA to institutionalize early forms of market-based environmentalism as alternatives to the allegedly heavy-handed forms of "command and control" (Shabecoff 1979; Layzer 2012, 78). Aptly capturing the zeitgeist, Lou Cannon of the *Washington Post* wrote, "The euphoria of the early Earth Days has been replaced by the hard realities of the budget squeeze, the conflict between the environment and the energy shortage, and the growing anxiety within industry against economic constraints" (Cannon 1973; cited by Layzer 2012, 40).

The problem, however, as EFFE's Kazis and Grossman argued, was that environmental regulations had relatively little impact on inflation; the cause was structural. As Kim Phillips-Fein notes, "Unions were making demands on employers that could no longer be met through increased economic productivity; business therefore passed the costs on to consumers through higher prices, which in turn sparked new demands for higher wages" (Phillips-Fein 2009, 156). From the end of World War II until 1973, productivity grew at an average of nearly 3 percent per year. From 1973 to 1979, it slowed to just over 1 percent (Bureau of Labor Statistics 2023). Economists Barry Bluestone and Bennett Harrison concurred with this assessment: corporations were able to abide the concessions to unions while profit rates were high during periods of sustained growth, but amid the tumult of the 1970s, union and environmentalist demands became intolerable (1982, 16–18). The corporate solution to the impasse? Disinvestment via plant closures and capital mobility, wrapped up in a rhetoric of job blackmail and a turn toward political organizing.

The Awakening of a Ruling Class-for-Itself

Lewis Powell's 1971 memo, written at the behest of his Chamber of Commerce neighbor, served as a rallying cry for corporate resentment that had been simmering while the state regulatory structure grew in response to the crises of the 1960s: air and water pollution, racial inequality, gender discrimination, multigenerational poverty, and occupational health and safety hazards. "It is essential," Powell wrote, "that spokesman for the enterprise system . . . be far more aggressive than in the past. . . . There should not be the slightest hesitation to press vigorously in all political arenas for support of the enterprise system" (Powell 1971). The corporate response to his call was swift and effective. Over the course of the decade, a whole network of conservative think tanks and trade groups was founded, including the Business Roundtable (1972), Heritage Foundation (1973), Cato Institute (1977), Reason Foundation (1978), and Competitive Enterprise Institute (1984). Those that had been around, like the Chamber of Commerce, took on a more aggressively partisan mission; a move that alienated some members but attracted many more (Phillips-Fein 2009, 177; Layzer 2012, 388). Membership in this new—avowedly conservative, militantly anti-statist—chamber skyrocketed from 100,000 in 1980 to 250,000 by 1982, and the power that it wielded in the halls of Congress grew as well (Crittenden 1982). Other long-standing organizations, like the American Enterprise Institute, found themselves awash with cash: "Between 1970 and 1980, the AEI's annual budget swelled from $1 million to more than $10 million" (Layzer 2012, 48). Corporate lobbying proceeded apace. David Vogel finds that "between 1968 and 1978, the number of corporations with public affairs offices in Washington increased from one hundred to more than five hundred. . . . The number of business-related PACs increased from 248 in 1974 to 1100 in 1978" (Vogel 1983, 31; citing Shabecoff 1979). These public affairs offices and PACs assembled lobbying campaigns that were remarkably effective in advancing corporate interests. Organized business successfully busted nearly the entirety of labor's legislative agenda: watering down the Full Employment Act to a largely symbolic measure and defeating both labor law reform and common situs picketing (Phillips-Fein 2009, 208–210).

The Chamber of Commerce, NAM, Business Roundtable and their allies also dramatically weakened both the Clean Air Act amendments and Toxic Substances Control Act; however, the continued bipartisan popularity of environmental laws created limits to the efficacy of legislative maneuvering on the environment. Beyond lobbying, capital's political flank pursued a variety of pathways to weaken existing laws and regulations, such as urging politicians to require the use of cost-benefit analysis and the incorporation of industrial expertise in environmental rule-making, and aggressively pursuing litigation aimed at challenging new environmental regulations.[2] But the most important struggle was waged in the court of public opinion. In the early 1970s, corporations began to devote much more money to advertising. Some of the biggest spenders on institutional advertising were fossil fuel companies: "Energy and utility companies were the most active. . . . The latter began devoting approximately half of their advertising budget to political persuasion" (Vogel 1983, 37). In one missive, Mobil Oil lambasted the Clean Air Act with the headline, "The $66 Billion Mistake" (Layzer 2012, 51). The American Electric Power Company took out ads in the *NY Times*, *Wall Street Journal*, and *Washington Post* challenging the "EPAs requirement that coal-fired utilities install scrubbers to remove SO2" (Layzer 2012, 51). The overarching message being communicated was that although some environmental laws were once necessary, the low-hanging fruit had been picked, and further environmental gains were more than offset by economic losses; businesses could be trusted to continue reforming their practices without the heavy hand of the state. The real environmentalists were to be found in industry (National Association of Manufacturers 1973b).

This political mobilization arose alongside four major changes to the mid-twentieth-century treadmill of production that I reviewed in chapter 2: the intensification of deindustrialization; the ascendance of financial capital; the growing power of transnational corporations amid globalization, and the accompanying decline of labor unions. First, the 1970s witnessed an escalation of the corporate-designed deindustrialization that had been building steam since the 1950s. Deindustrialization proceeded through many paths; companies could

> reallocate profits earned from one plant's operations to new facilities, . . . reallocate capital by running down a plant simply by failing to replace worn out or obsolete machinery, . . . physically relocating some of the equipment from one facility to another or selling off some of the old establishment's capital stock to specialized jobbers, . . . [or] completely shutting down a plant. (Bluestone and Harrison 1982, 7)

In their seminal 1982 work, *The Deindustrialization of America*, economists Bluestone and Harrison found that "somewhere between 32 and 38 million jobs were lost during the 1970s as a result of private disinvestment in American business" (1982, 9).

Second, deindustrialization was accompanied by broader changes in the organization of jobs and the creation of profits. Jobs in the heavily unionized manufacturing sector plummeted, while those in the largely nonunionized service sector skyrocketed. Beginning in the 1980s, however, profits were increasingly concentrated in neither of these industries but in the finance, insurance, and real estate (FIRE) sector (Krippner 2005, 177–180; see also Strange 1986; Arrighi 1994; Stein 2010). Jeffrey Frieden notes that "international financial markets grew from $160 billion in 1973 to $3 trillion in 1985" (2006, 397). The impacts of financialization rippled through the economy: "Confronted with labor militancy at home and increased international competition abroad, non-financial firms responded to falling returns on investment by withdrawing capital from production and diverting it to financial markets" (Krippner 2005, 182).

Third, this "financial fix" (Silver 2003) was paired with what geographer David Harvey refers to as a "spatial fix"; as corporations organized across increasingly global chains of production, they sought out lower production costs not only by shifting operations to domestic hinterlands (e.g., right-to-work states in the US South), but to developing countries. Bluestone and Harrison find that "between 1950 and 1980, direct foreign investment by U.S. businesses increased sixteen times, from about $12 billion to $192 billion. . . . By the end of the 1970s, overseas profits accounted for a third or more of the overall profits of the hundred largest multinational producers and banks in the United States" (1982, 42). According to Frieden, "From the early 1970s to the early 1980s new foreign investment by multinational corporations soared from about $15 billion to nearly $100 billion a year" (2006, 397). This trend would further accelerate over the course of the 1980s and 1990s (Burke, Epstein, and Choi 2004).

Fourth, by the dawn of the 1980s, labor unions had "been forced into a defensive posture," seeking to hold on to what they'd previously gained rather than asking for more, and even being forced into concessions in many cases (Kazis and Grossman 1982, 95). Writing in *Dissent*, Robert Kuttner reported that,

> in the 1980s, new union organizing brings in only 100,000 to 200,000 private-sector members a year, compared to 750,000 a year in the 1950s. In 1984, according to the AFL-CIO, the labor movement organized about 150,000 new private-sector workers, and about 200,000 in the public sector. But in the same year the labor movement lost about 700,000 workers from plant closings, automation, layoffs, and decertification elections, leaving a net loss of 350,000, or about 2 percent of its total membership. (1984, 53)

Such a conjuncture opened up space for a new alliance between capital and *disorganized* labor; a shared political interest between business and both alienated rank-and-file unionists and nonunion segments of the working class.

Amid the political economic maelstrom, it was easy for capital to rely on the tactic of "job blackmail," which was used not only to attack potential laws and regulations but also to place the blame on environmentalists for any job losses or plant closures that did occur. In *Fear at Work: Job Blackmail, Labor and the Environment* (1982), Kazis and Grossman catalogue examples of this strategy in action. For example, when Anaconda Copper closed its smelter in Anaconda, Montana, and its refinery in Great Falls, Montana, they announced at a press conference that "the company has determined by in-depth studies that the existing plan cannot be retrofitted to satisfy environmental standards" (Kazis and Grossman 1982, 3). The plant's 1,500 employees would be out of a job. Workers met with the governor, carrying signs that read "Our Babies Can't Eat Clean Air" (Kazis and Grossman 1982, 4). Similarly, when the EPA ordered Union Carbide Company to comply with new air pollution standards in Marietta, Ohio, "the company announced that it could reduce the plant's sulfur dioxide only by shutting down two boilers and laying off over 600 workers" (Kazis and Grossman 1982, 7). In both instances, the companies were lying. In the case of Anaconda, a local columnist captured the crux of the matter: "Someone long ago in the Anaconda hierarchy made the decision not to upgrade the antiquated smelter and it was allowed to deteriorate" (Kazis and Grossman 1982, 6). In the case of Union Carbide, the OCAW protested the decision, and ultimately "the plant switched to low-sulfur coal, met the pollution-control deadline, and did not fire a single worker" (Kazis and Grossman 1982, 11).

Truth be damned, such proclamations from industry were all the proof that politicians needed to jump into action. In 1971, a Department of Commerce official warned that "environmental regulations would cause such severe economic dislocation and unemployment that major new relief programs would have to be created" (Kazis and Grossman 1982, 17). The Nixon administration's solution was to have the EPA implement an Economic Dislocation Early Warning System (EDEWS) to track and assess the impact of environmental regulations on plant closures and job losses. Murray Weidenbaum, an economist working with the American Enterprise Institute, published several reports analyzing the purported economic impact of "over-regulation." Expressing particular concern about OSHA and the EPA, he observed that "when the clean air and clean water act amendments of 1977 reach their full impact, it will be extremely difficult to build a factory in many parts of the United States" (1980). In subsequent studies, however, the EPA, Bureau of Economic Analysis, and unions themselves found the job-related losses from environmental regulations to be minimal (Oil, Chemical and Atomic Workers International Union 1976, 10; Kieschnick 1978, 26; US Environmental Protection Agency 1983). A representative for the steelworkers underscored this point: "As far as we're concerned in the Steelworkers, we don't know of any single facility that had to shut down because of environmental clean-up" (Kazis and Grossman 1982, 21). Based on these studies, as well as their own analysis, Kazis and Grossman concluded, "Employers rarely close a facility solely because of pollution control requirements. In most cases, obsolescence, declining sales, problems with raw materials, more efficient competitors, and increased energy costs are much more important" (1982, 19).

In spite of the fact that there existed little evidence that environmental regulations were leading to widespread job loss, the narrative of "jobs versus environmental protection" stuck. Moreover, while groups like Environmentalists for Full Employment worked to call attention to the plight of labor, the fact that many greens—both mainstream and radical—failed to see the connections reinforced the sense among some rank-and-file workers that environmentalists were out of touch with the needs of the working class. On its own, corporate political organization had an impact on the prospects, substance, and implementation of environmental laws; paired with job blackmail, this new wave of corporation mobilization sent a shockwave through the whole of US politics. Shabecoff (1979) characterized this as a "new cohesiveness and assertiveness of the business community." More bluntly, "the business community became more class conscious" (Vogel 1983, 34).

THE NEOLIBERAL CLASS ECOLOGICAL ORDER

The political economic crises of the 1970s created what public policy scholars refer to as a "punctuated equilibrium"; a conjuncture in which there exists an incongruity (perceived or real) between entrenched ideologies and policy responses and the scale and scope of existing problems. The result is that space opens up for new ways of intervening in the crises of the moment. In this case, the ideology that surged to the fore was not new, but had been marinating in corporate boardrooms and conservative think tanks for nearly forty years: excessive state intervention was leading us down a road to serfdom, and in the free-market lay the roadmap for social and ecological flourishing. The concrete interventions favored by neoliberals—austerity, deregulation, privatization of core state functions, and rampant tax cuts—found their legs in the authoritarian palace of Pinochet and the undemocratic halls of the Bretton Woods institutions. However, in the United States, past attempts to advance the political vision of neoliberalism had been stymied by the transparently inequitable impacts that the proposed policies would have on the US masses; a more sophisticated public relations strategy that appealed to the "common man" was required.

The conservative embrace of the extractive laborer is nonetheless more paradoxical than it might initially appear: conservatives turned to the working class at the very moment when its obituary was being written the world over, with its most persuasive authors coming from the political Left. Over the course of the 1970s, Cowie remarked, "one of the great constructs of the modern age, the unified notion of a 'working-class' crumbled, and the new world order was built on the rubble" (2010, 18). Amid the smoldering ashes of deindustrialization, philosopher André Gorz (1982) argued that the real revolutionary subject was not the working class but "a non-class of non-workers" who could lead us to a world beyond wage labor. By the mid-1980s, sociologist Ulrich Beck could confidently declare that "the notion of a class society remains useful only as an image of the past" (1992 [1986], 91).

Such apparently radical musings meshed surprisingly well with the reformist liberalism of the era. For example, political scientist Ronald Inglehart made the case that recent years had witnessed a profound "political realignment in advanced industrial society: from class-based politics to quality of life politics" (Inglehart and Rabier 1986). This quality of life politics, for Inglehart and his collaborators, was an indication of a turn toward "post-material values"—"self-expression, belonging and quality of the physical and social environment"—that was enabled by industrial affluence and particularly apparent in the middle-class suburbs discussed in chapter 2. In efforts to attract this quality-of-life voter, over the course of the 1980s, the so-called New Democrats advanced a vision of the "Third Way"—an alternative path, beyond traditional Left/Right distinctions, which eschewed "the political ideas and passions of the 1930s and 1960s," and embraced "the free-market" . . . "as the best engine of general prosperity" (Democratic Leadership Council 1990). In this framework, the Left's historical focus on class conflict was eclipsed by calls for consensus, entreaties to entrepreneurship, praise for public-private partnerships, and an unbridled confidence that economic growth could smooth over social and political antagonisms. In the quest for late-twentieth-century prosperity, the new liberal of the 1980s was "prepared to leave the mechanism of the New Deal behind" (Rothenberg 1984, 27).

In spite of continued union-bashing and ritualistic homages to the beneficence of corporate tycoons, it was elements of the New Right who, outside of labor unions and the marginal voices that remained of the old Left, voiced some of the era's most strident expressions of anger at the plight of US workers. Building on tropes of the "Silent Majority" and "Middle-America" discussed in chapter 2, conservatives strengthened their appeals to the white working class in a renewed attempt to create a racialized, cross-class coalition with the ruling class, one in which white male heavy-industrial and extractive sector workers occupied the center stage. This political strategy was informed and intensified by the calculations of capital. In response to threats posed by labor and environmentalism amid declining rates of productivity and profit, and sensing a political opportunity amid the malaise of the 1970s, organized business surged to the fore, leaving the long-precarious New Deal political coalition in tatters. Anti-environmentalism was not only integral to the building of this neoliberal world order; it was the key to a new conservative conception of class politics—a class struggle from above—that has been waged ever since, with quite literally earth-shattering results.

So began a several-decade-long path toward the hyper-politicization of environmental issues. However, this hyper-politicization didn't come naturally; clean air and water, wilderness and open spaces, and protection from toxic substances remained political goods that most constituents strongly desired. And the obvious economic interests of the corporate lobby in privileging profits over environmental protection put limits on the persuasive capacity of business itself. Only by embedding corporate opposition to environmental regulation in a broader cultural narrative would anti-environmentalism become a defining element of US conservatism. More to the point, it was only by defining environmentalism and anti-environmentalism in class terms that conservatives began to have sustained success in advancing their market fundamentalist visions and retrograde cultural dreams (Fraser 2016). There were three actors who constituted the conservative, anti-environmentalist vision of class struggle: the environmentalist as New Class elitist; the (poor, black and brown) underclass as a threat to social order; and the (white male) extractive laborer as the vanguard of the working class.

The Environmentalist as New Class Elitist

In her 1989 book, *Fear of Falling*, Barbara Ehrenreich noted that "in the intellectual theories of the right, the liberal elite is much more than a political antagonist. It is a class, known normally as the New Class." The origins of the New Class can be traced back to several mid-twentieth-century works by socialist Max Shachtman, leftist-turned-conservative James Burnham, and Yugoslav dissident Milovan Djilas—all of whom reflected on the power of a bureaucratic elite in Communist states, and the failure of those in such positions to act in the interests of the working-class masses. It wasn't until the 1970s, however, that the concept picked up steam in conservative discourse (see, e.g., Novak 1973; Hart 1975; Kristol 1975; Bruce-Briggs 1979; Bazelon 1979; Gouldner 1979; Podhoretz 1979). The basic idea was that, alongside the rise of the postwar welfare state, a class of well-educated professionals had ensconced itself within universities, think tanks, and government agencies and was using its expertise—rather than its economic clout—as a means to advance its own interests. Definitions of the New Class varied; some analysts emphasized its "secular hedonism" (Novak 1973, Kemp 1982), others its adoption of an "adversary culture" (Podhoretz 1979), still others its participation in bureaucratic planning (Bazelon 1979), and others its status and wealth relative to the working class (Ladd 1979). In some discussions, the New Class seemed to cohere around nothing more than an "arbitrary selection[s] of people who are professionals as well as liberals" (Ehrenreich 1989, 152).

In virtually all these accounts, both the grassroots activists of the environmental movement and the bureaucrats charged with implementing environmental regulations were construed as prototypical new class elitists. Journalist Jeffrey Hart (1975) wrote that adherents to the New Class are "opposed to the goals and achievements of liberal capitalism. . . . They despise economic growth as ecologically unsound, but also—and more importantly—as morally reprehensible." In such accounts, the New Class not only used environmental laws and regulations to advance its own moralistic vision for America, but to shore up its power. According to *Wall Street Journal* editor Robert Bartley, the intentions of the New Class "are quite clear: a society in which rewards would no longer be in wealth, but in power and status, to be won by precisely those skills (abstraction, moralistic rhetoric, manipulation of symbols) in which the highly educated New Class excels" (1979, 59).

The New Class, in this telling, stood opposed to two forces: capital and the working class. Pat Buchanan, for instance, argued that "the businessman is engaged in a class war with upper-middle-income liberals for authority and social position." Most people, he explained, "are contemptuous . . . of the environmental extremists more alarmed over Atlantic fish and Alaskan caribou than the jobs of autoworkers or the economic independence of the United States" (Buchanan 1976). A 1978 article in *Conservative Digest* similarly asserted that "it is a liberal who, by his actions, is creating a paradise for wildlife at the expense of the jobs and improvement in man's material welfare" (Crane 1978). Noting the gap between union leadership and rank and file, Richard Viguerie (1977) commented that "members of the ultra-liberal United Auto Workers recently joined with auto manufacturers in fighting the ecology extremists on auto emissions standards and won an important victory." He then listed issues of common ground between union members and conservatives: "jobs, crime, government interference in individual lives, ecology extremism and busing." "At heart," journalist William Tucker observed, "environmentalism favors the affluent over the poor, the haves over the have-nots" (1982, 36).

Broadsides against the New Class provided conservatives an opening to appeal to the proverbial common man while waging war on unions, environmentalists, and the welfare state. The principal victims of this war, however, were not environmentalists or bureaucrats but the demographic grouping that had come to be known as "the underclass." "With the ascent of the New Right in the eighties," Ehrenreich observes, "the idea of the New Class emerged from the airy realm of thought to become part of the rationale for actual policy: policies directed, however, not against the New Class or anything resembling it, but against the poor" (1989, 161).

The New Class and the Underclass

In the 1970s, a variety of conservative commentators began describing a new social terrain, where New Class elitists and state bureaucrats attempted to ally with the poor and racial minorities in opposition to business and the working class. "Liberals," wrote Tucker, "are the upper-middle class people defending the rights of the poor, while conservatives are the lower-middle class people defending the rights of the rich" (Tucker 1982, 28). This New Class alliance with the poor was, according to conservatives, an instrumental and opportunistic one; as one *National Review* editorial put it,

> The poor must be understood in a special sense, as potential clients for the redistributive ministrations of the New Class, the middlemen of social justice. Analytically, "the poor," as a concept, legitimates the power-grab of the New Class middlemen in the same sense as "the proletariat" legitimates the power-grab of the Leninists. (Ehrenreich 1989, 185)

William Rusher concluded that the "basic cleavage in American politics today" was a "new division [that] pits Establishment WASPs plus their minority group allies against middle Americans and the hyphenated-ethnics (Italo-Americans, Polish-Americans)." This alliance between the New Class and poor racial minorities was, according to Rusher, "a new patronage network": "Establishment WASPs are in a good position to 'pay off' their minority group allies with all sorts of cultural and economic goodies" (1975, 36).

Beginning in the late 1970s and intensifying over the coming decade, these poor, minority group allies came to be referred to by politicians, the media, and scholars alike as "the underclass." Perhaps even more so than the New Class, the underclass lacked any sort of coherent definition. Instead, as political scientist Adolph Reed noted, narratives invoking the concept "appeal to hoary prejudices of race, gender, and class that give the underclass image instant popularity and verisimilitude even though it is ambiguous and inconsistent on its own terms" (Reed 1999, 179). As "a concept in search of an object" (Reed 1999), the underclass nonetheless came to color both conservative and mainstream liberal political discourse. For Charles Murray, the underclass referred to the "growing number of poor people engaged in self-destructive personal behavior that [will] keep them at the bottom of society" (1984, xvi; see also Murray 1990). Writing in the *New Yorker*, Ken Auletta broke the underclass into four groups: "hostile street criminals, hustlers, mothers, and the traumatized." "Members of the underclass," he wrote, "are responsible for a disproportionate amount of the crime, the welfare costs, the unemployment, and the hostility that beset many American communities" (1981,105). In these portrayals, the underclass were jobless, dependent on government aid, from broken families, and engaged in criminal practices. A 1977 *Time* magazine article warned that

> behind the [ghetto's] crumbling walls lives a large group of people who are more intractable, more socially alien and more hostile than almost anyone had imagined. They are the untouchables: the American underclass. . . . Their bleak environment nurtures values that are often at odds with those of the majority—even the majority of the poor.

Some analyses of the underclass focused explicitly on poor African Americans in inner cities, while others included whites; some defined the underclass in terms of poverty, while others noted that a portion of those working in the informal economy were not poor. One area of consistency in these analyses was that the underclass was defined not by its members' relationship to capital but by their refusal to forge a relationship to capital. And yet, building on Daniel Patrick Moynihan's famous report and Oscar Lewis's analyses of the "culture of poverty," conservative accounts of the underclass served to steer attention away from the political economic forces that produced poverty and toward cultural explanations for its persistence in particular communities. The underclass was struggling economically, lived in the ghetto, and had adopted attitudes and behaviors outside of the middle-class mainstream; for the underclass, the welfare state had produced a "welfare culture." According to Rusher, "Most of the values on which social and economic conservatives agree have come under sustained attack from the liberal verbalist elite, supported politically and psychologically by its own proudest creation: an entire new economic and political constituency geared to the welfare ethic" (Rusher 1975, 153). It was precisely this reliance on "welfare culture" that, to *New Republic* editor Mickey Kaus (1990), defined the underclass. As Pat Buchanan mused, "Poverty is not simply an economic condition. It is often the consequence of a particular perspective on life that is not changed by changing one's zip code. Slum people taken out of slums, put into public housing, have quickly recreated the environment they recently left" (1976). With such characterizations in mind, Reed astutely observed that "the underclass idea's power derives from its naturalization of 'culture' as an independent force that undermines adaptability and retards progress" (1999, 184).

The underclass lacked any real-world sociological referent—no one, as Reed pointed out, thought of themselves as a part of the underclass—but it nonetheless surged to the fore of policy responses advanced by conservatives and, in some cases, embraced by liberals. Reagan continually peppered his speeches with references to "welfare queens," fears of the Black "gangster" legitimated a rapid expansion of the carceral state, and—soon enough—the "illegal alien" was used to further attacks on social programs and propel border militarization. As sociologist Loïc Wacquant contends, the underclass "started out as a proto-concept . . . but quickly morphed into an instrument of public accusation and symbolic disciplining of the threatening black precariat in the hyperghetto" (2022, 2).

The reality was these pockets of intense urban poverty were products of a number of intersecting forces: (1) the aforementioned deindustrialization that had plagued cities (particularly in the Northeast and Midwest) since the 1950s; (2) the historical legacies of the intense institutional racism that had long accompanied both hiring and housing policies; and (3) an increasingly deregulated housing market whose patterns of racial discrimination shifted from overt exclusion to a form of "predatory inclusion" in which "housing value in the United States continued to be scaled according to the proximity of African Americans" (Taylor 2019, 254). This resulted in a population of poor and working-class people, disproportionately African Americans, confined to urban centers that had been decimated by the flight of both capital and their more economically secure neighbors, deprived of the tax base needed to provide public services, and caught in the cross-hairs of a government rapidly shifting away from the provision of a social safety net and toward the criminalization of poverty. The size of this shift is striking:

> In 1980, the country spent three times as much on its two main assistance programs ($11 billion for Aid to Families with Dependent Children [afdc] and $10 billion for food stamps) than on corrections ($7 billion). By 1996 . . . the carceral budget came to double the sums allocated to either afdc or food stamps ($54 billion compared to $20 billion and $27 billion, respectively). (Wacquant 2010, 76–77)

A federal jobs program with a goal of providing full employment at a living wage—like that envisioned by advocates of Humphrey-Hawkins—would have been one logical and equitable solution to the crises created by deindustrialization and globalization; instead, the United States turned to what geographer Ruth Wilson-Gilmore calls "the prison fix." According to Wilson-Gilmore, the construction of new prisons was an attempt to absorb surpluses—of land, labor, state capacity, and finance capital—that had emerged out of the aforementioned political economic ruptures. The end result of this was the hyperincarceration of impoverished, young, African American men from the inner cities.[3]

The prison fix, coupled with the massive curtailment of social services, intensified inequality and created enormous suffering. To conservatives, however, those dependent on the welfare state were non-productive, irresponsible threats to society who were incapable of exercising political agency—mere pawns in the self-interested games of New Class elitists. From the perspective of this worldview, as Ehrenreich observed, "the New Class alliance with the poor and minority groups left an obvious strategic alliance for the Republicans—with the working-class" (Ehrenreich 1989, 165). In a tragic twist of fate, the Republican shift in attention to this demographic occurred at the very moment that the Democrats were turning their backs on the working class.

The New Democrats and the Decline of the Working Class

Amid the tumult of the late 1970s and the loss of the presidency and Senate (for the first time since 1952) in the 1980 election, the New Democrat emerged. Growing out of the Coalition for a Democratic Majority and House Democratic Caucus' Committee on Party Effectiveness, and eventually associated with the Democratic Leadership Council, New Democrats were an amalgamation of liberal cold warriors, Southern centrists, and politicos alienated by the perceived overreach of the New Left (Hale 1995; Baer 2000).[4] But the faction's ideological lodestar was the "new liberalism" associated with a younger generation of Democratic politicians, like Gary Hart, Al Gore, Bill Bradley, Timothy Wirth, Richard Gephardt, Bruce Babbitt, and Paul Tsongas, as well as a coterie of journalists and scholars such as Charles Peters of the *Washington Monthly* and the MIT economist Lester Thurow (Rothenberg 1984; Peters and Keisling 1985). The mantra of the new liberals was that "the solutions of the thirties [were] not going to solve the problems of the eighties." This emergent ideology was an explicit rejection of New Deal politics; as Randall Rothenberg put it in his flattering book-length portrayal,

> in the social arena, redistribution gave way to investment. In economics, macroeconomic policy was accompanied by a fascination with microeconomic matters. . . . The industrial age had given way to the information era. And with it, liberalism was overtaken by neoliberalism. (1984, 18)

For Peters, the new liberalism represented a kind of "liberal realism"—"we criticize liberalism not to destroy it but to renew it by freeing it from its myths, from its old automatic responses in favor of unions and big government and against business and the military" (Peters and Keisling 1985, 10).

Economist Pat Choate similarly argued that these leaders were advancing a politics grounded in "pragmatism, of breaking past ideological boundaries to determine how to restore long-term economic growth, and to do it in a socially responsible and equitable manner" (Rothenberg 1984, 25).

In truth, this was a political rationale that was born out of the ashes of defeat, an effort by a segment of the Democratic Party to respond to the triumvirate of deindustrialization, oil shocks, and stagflation in the aftermath of McGovern's loss and Reagan's ascendance. For the New Democrats, state intervention had emerged as the de facto response to any problem—poverty, labor relations, the environment, and on—but, as a consequence of bureaucratic hubris and excess, often failed to ameliorate these problems. The ideology of social liberalism that had brought us the New Deal and Great Society had lost its capacity for self-reflection. The solution to these problems lay in a "third way"—the use of "appropriate technology" and public-private partnerships, guided by the knowledge of cost-benefit analysis, an almost religious faith in the ability of economic growth to dampen social antagonisms, and a concomitant rejection of concern over ecological limits that had characterized (and energized) liberal activists of the 1970s. At a moment when the New Deal coalition was crumbling—and working-class people were losing their jobs and houses—the New Democrats turned their attention away from organized labor and toward the median voter, away from demand and redistribution and toward supply and investment, away from class relations and toward the individual entrepreneur, away from political conflict and toward consensus. Regardless of the intellectual merit of their project, the New Democrats ruled the day: their philosophy was endorsed by a stable of journalists and public intellectuals, institutionalized in the Democratic Leadership Council, and—after Bill Clinton's 1992 victory—solidified as the governing philosophy of the Democratic National Committee. As the Democratic Leadership Council put it in its 1992 Hyde Park Declaration, "We believe in a Third Way that rejects the old left-right debate and affirms America's basic bargain: opportunity for all, responsibility from all, and community of all."

What did such a vision herald for workers and environmentalists? With regard to labor, the New Democrats were critical of what they alleged to be union intransigence in the face of the economic malaise of the late 1970s and early 1980s (e.g., refusals to accept wage freezes) and unflinching Democratic support for unions that had grown bureaucratic and bloated (Peters 1982). In this line of thinking, a political economic philosophy founded on an antagonistic relationship between capital and labor obscured the greater public interest and the possibility of cooperation in the workplace (Rothenberg 1984, 47). "The New Democrats," political scientist Jon Hale recognized, were "more closely tied to business than to organized labor and take pains to distance themselves from Jesse Jackson's Rainbow Coalition" (1995, 224). They were particularly hostile to labor's desire for economic protections for US workers in a rapidly globalizing world: "The question for the Democrats is, will they accede to labor's desire to protect jobs from the vagaries of international competition, or will they accept the new reality and the consequences of a global economy?" (Rothenberg 1984, 91). They chose the latter. With former DLC-leader Clinton in the presidency, the party embraced so-called free trade with a missionary-like zeal; a position that met opposition from unions and most environmental organizations (though, in 1993, when the Clinton administration lobbied for the North American Free Trade Agreement, it found support from the Environmental Defense Fund, Nature Conservancy, and National Wildlife Federation) (Baer 2000, 184–187).

For the New Dems, the twenty-first-century US economy lay not in the unionized workers of the manufacturing sector but in the knowledge workers in high-tech "information archipelagos." As Dick Gephart put it in 1982, "The future of the American economy is in having the best thoughts, the best mental work, as opposed to having a work force that is particularly adept at making things" (Rothenberg 1984, 85). Rather than propping up the declining industrial sector through state intervention, the logic went, we should focus attention on stimulating emerging growth areas: independent entrepreneurs and the technology and service sectors (1984, 76–77). "Our hero," read Peters's "Neo-Liberal's Manifesto," "is the risk-taking entrepreneur who creates new jobs and better products."

This emphasis on entrepreneurship and markets extended into environmental policymaking. New Democrats rejected the position of some environmentalists that economic growth was a danger to the environment—a line of thinking that blossomed during a period of sustained growth but was less convincing at a moment of economic hardship (Rothenberg 1984, 65–66). Accompanying this emphasis on growth, neoliberals advocated market-based forms of environmental protection as an alternative to the supposed heavy hand of "command and control" regulations, such as the Clean Air and Water Acts of the prior decade. In a 1986 op-ed in the *Wall Street Journal*, Fred Krupp, of the Environmental Defense Fund, argued that a new wave of environmentalism was forging ties with industry and focusing on the mutually beneficial gains that could accompany environmental regulation. The Progressive Policy Institute made the case for more "market-based, information-driven, and community-friendly ways to protect the environment" (Mazurek 2003). Such instruments had, in fact, long been integrated into the supposed command and control structure of US environmental regulations (Portney and Stavins 1998, 8), often due to successful industrial lobbying, but they were now emerging as the organizing principle that would shape subsequent environmental policy proposals over the next several decades.

This neoliberal vison for environmental policy was predicated on the acceptance of the very conservative talking points that groups like Environmentalists for Full Employment had fought so hard to combat: that, absent capitulation to capital's demands, there exists a real trade-off between jobs and environmental protection, that "command and control" environmental laws and regulations are too expensive, and that state regulations are innately inefficient. Rather than bridging the gap between jobs and environmental protection by working alongside unions, the preference was to work toward solutions long advocated by capital. As Kathleen McGinty, head of the White House Office of Environmental Policy under Clinton, put it, "It's time for eco-warriors and industry to realize that their interests don't make them mortal enemies" (2004). Perhaps this is the reason why the Koch Corporation was among the twenty-eight corporations who once made up the DLC's Executive Council (Dreyfuss 2001, 22). The path to mutually beneficial gains was paved by the creation of markets in which permits to pollute could be traded: in greenhouse gases, ozone depleting substances, stationary sources of air pollution, and so on. "Command and control" approaches, the narrative went, produced important benefits but "sometimes stifled innovation by locking firms into a specific kind of equipment and increased regulatory costs and burdens by taking such a detailed, inflexible approach" (Clinton 1992).

This turn toward market-based environmental regulations was in part strategic: there was hope that such a move could divide industry (by enticing those companies with first-mover advantages to support market-based policies), keep Democrats relevant in an increasingly Republican-dominated South, win the support of Republican Party members opposed to the regulatory state, and appeal to the median voter—always envisioned as a middle-class suburbanite—skeptical of state spending (Layzer 2012, 414–441; Aronoff 2021, 73–79). Over the course of the 1990s—with the modest success of the sulfur dioxide permit system[5] and the failure of Clinton's BTU tax proposal (discussed in the section that follows)—cap and trade became the preferred vehicle to tackle climate change as well. The New Democrats laudably recognized the necessity of responding to climate change, but lacked the political savvy to turn this recognition into a viable path toward policy reform.

Like the social policies embraced by the New Democrats, discussions of cap-and-trade systems and carbon taxes were hitched to the policy preferences of a median voter and guided by the expertise of policy wonks and technocrats. In other words, at a time where the Republicans were setting their sights on the New Class elitists, Democrats doubled down and embraced the New Class as the path toward progress; the model US citizens of the information age. Labor was downgraded to a bit concern, while for the unemployed, those reliant on social welfare, racial minorities living in inner-cities, and undocumented immigrants, the ideology advanced by the Clinton-era Democratic Party was particularly pernicious. Echoing the conservative narrative that I outlined in the previous section, the Progressive Policy Institute published a report titled the "Politics of Evasion," musing that the Democratic Party was "increasingly dominated by minority groups and white elites—a coalition viewed by the middle class as unsympathetic to its interests and its values" (Galston and Kamarck 1989, 4). The unequivocal message here was that the party needed to move to the center in order to appeal to the suburban middle class.

These middle-class suburban voters, it seemed, were an anxious bunch who embraced much of an emergent bipartisan narrative about the harsh realities of the late-twentieth-century world and the nature of those being left behind. Under the Clinton administration, the Democratic party gutted the social welfare state, while ramping up the carceral state; ushered in a new era of "free trade" that wreaked havoc on the working class and the environment, at the same time as they militarized the US-Mexico border; and deregulated finance and telecommunications, while leaving organized labor in tatters. When Al Gore lost the 2000 election, the DLC's Al From doubled down on the New Democratic solution, blaming Gore's defeat on his appeals to manufacturing workers: "By emphasizing class warfare . . . he seemed to be talking to Industrial Age America, not Information Age America" (Dreyfuss 2001, 24). The sentiment that the working class was a figment of the past was echoed by an editor of the DLC's *Blueprint* magazine: "You can't have class warfare without classes. . . . All these guys have boats in their backyards" (Dreyfuss 2001, 24).

THE EXTRACTIVE LABORER AS WORKING CLASS

Republicans responded to this Democratic shift by loudly proclaiming—in the midst of a massive project of union-busting, deregulation, and privatization—that *they* were the party of the US worker. But who was the working class in this emergent conservative vision? The public statements made by conservatives were overwhelmingly structured around white male workers in the extractive sector and, to a lesser extent, chemical and energy-intensive forms of heavy industry. Environmental politics—specifically, debates over public lands, logging in the Pacific Northwest, energy production, and climate change—played a crucial role in the class struggle being waged by conservatives. This was particularly apparent in, and dependent on, the material and symbolic politics of the American west (Turner 2009). Over the course of Reagan's term in office, the grassroots anger of the Sagebrush Rebellion and the machinations of corporate-funded conservative think tanks coalesced into the Wise Use movement. While Wise Use shared Sagebrush's goals of devolving control over public lands to western states, its broader agenda "outlin[ed] a far more diverse and sophisticated set of demands" (Cawley 1993, 166), which read like a wish list for the National Association of Manufacturers and Chamber of Commerce: development of oil and gas resources in the Arctic National Wildlife Refuge; opening all public lands to mineral and energy production; devolving control over all waters to the states; gutting the National Park Service; weakening the Endangered Species Act; and intensifying timber extraction in the name of responding to climate change (Arnold and Gottlieb 1988, Arnold 1993).

Wise Use was nonetheless novel in terms of the scope of the constituencies that it mobilized and the nuance of its narrative. Wise Use comprised a loose network of local workers and community members, joined together by national organizations—like Alliance for America, People for the West!, and the Yellow and Blue Ribbon Coalitions—who received ample support from conservative think tanks (American Legislative Exchange Council, American Freedom Coalition), legal aid foundations (Mountain States Legal Foundation), antiregulatory organizations (e.g., Center for the Defense of Free Enterprise), and industrial lobbies (American Farm Bureau Federation, Cattleman's Association, American Mining Congress, Petroleum Institute, California Forestry Association, etc.) (Brick 1995, 19). It combined the grassroots rage of the Sagebrush Rebellion with the legal logics and libertarian ideologies of the property rights movement and the institutional muscle of the Republican Party (Helvarg 1994, 123). In spite of the movement's vehemently anti-statist posture, Wise Use advocates positioned themselves as representing "a new balance . . . a middle way between extreme environmentalism and extreme industrialism" (Arnold and Gottlieb 1988, ix; see also Watt 1982; Arnold 1993). The concept of "wise use" was itself appropriated from the conservationist Gifford Pinchot, who once remarked that "conservation means the wise use of the earth and its resources for the lasting good of men." While advanced and encouraged by corporate executives, this narrative strategy was parroted by miners, ranchers, and lumberjacks who continually caught the eye of the media.

Nowhere was this more apparent than in the struggles over the preservation of old-growth forests in the Pacific Northwest. On one side of these struggles were environmental groups who sought to preserve old-growth forests and protect endangered species. Radical greens, like members of Earth First!, pursued direct action that included mass protests, tree-sits, tree-spiking, chaining themselves to logging equipment, and blockading logging roads in an effort to defend the "ancient forest." Mainstream environmentalists, like the Audubon Society and Sierra Club, turned to the courts in efforts to compel federal agencies to abide by the stipulations of the Endangered Species Act (particularly with regard to the northern spotted owl). On the other side were timber companies, large and small, and the workers and communities whose livelihoods depended on industrial forestry. They organized, in groups like the Oregon Lands Coalition, to call attention to the impact that limitations on logging would have on communities that were already hemorrhaging jobs and a tax base. At one rally in Forks, Washington, angry community members carried signs with messages like "No timber, no revenue, no schools, no jobs" and "Don't take my daddy's job" (cited by Layzer 2016, 250). An advertisement placed by the Yellow Ribbon Coalition in the magazine *American Timberman and Trucker* read, "Hard-working men and women—family people—are losing their jobs, the direct result of preservationist lawsuits and timber sale appeals that are locking up our forests. We must unite in our common defense" (Switzer 1997, 191). Wise Use leader Ron Arnold accused environmentalists of "putting rats ahead of family wage jobs" (Arnold 1993; see also Brick 1995, 39). Evelyn Badger, of the Oregon Lands Coalition, urged Congress to care about endangered communities as much as they care about endangered species (Badger 1992; cited by Turner and Isenberg 2018, 85).

For Wise Use activists, the real environmentalists were "the farmers and ranchers who have been stewards of the land for generations, the miners and loggers and oil drillers who built our civilization by working in the environment every day, the property owners and technicians and professionals who provided all the material basis of our existence" (Layzer 2012, 173). Environmentalists, on the other hand, were—as Arnold put it—"part of an elite, part of the Harvard Yale crowd in three-piece suits and expensive shoes that is destroying the middle class" (quoted by Layzer 2012). Environmentalists, at times, played into the Right's caricatures by failing to consider the plight of workers or even mistaking extractive workers for their political opponents. When a mill worker was almost killed after his saw hit a log that had been spiked, Dave Foreman of Earth First! suggested that while "it's unfortunate that somebody got hurt, . . . I quite honestly am more concerned about old-growth forests, spotted owls and wolverines and salmon—and nobody is forcing people to cut trees" (Bari 1994, 268; cited by Loomis 2012, 114). The Native Forest Council mused that "companies and industries have been changing or shutting down for 200 years, and workers always find new jobs. . . . Chopping down forests for the sake of jobs is nothing more than social welfare—not something our nation prides itself on" (Foster 2002 [1993], 119).

The reality was that the plight of workers and communities in the Pacific Northwest had little to do with spotted owls or environmental activism. Public lands factored into this debate only because private capital had already cleared virtually all of the fully intact old-growth forest stands from its lands, leaving smaller timber companies in a precarious position; "lacking private forest lands of their own," they were forced to "rely almost entirely on access to public tim- ber to feed their mills" (Foster 2002, 114). As John Bellamy Foster notes, the structural origins of the plight of logging towns lay in the "conditions of eco- nomic stagnation" of the era: a precipitous decline in housing starts, increased competition from foreign logging companies, and increased demand for US timber supplies abroad (2002, 116). This placed particular stress on the high- value old-growth timber in the Pacific Northwest. Timber companies were receiving sweetheart deals from the federal government—including a bailout amid the housing crash of 1982 and tax exemptions for export income begin- ning in 1984—at the same time as they were (1) demanding wage and benefit cuts from their workers, (2) investing heavily in mechanized equipment that required fewer workers, and (3) shifting operations from the Northwest to the Southeast in pursuit of lower labor costs (Foster 2002, 121).

Regardless of their merits, these anti-environmental arguments resonated far beyond the confines of timber country and had some success in shifting societal perceptions in communities dependent on industrial-scale resource extraction. Helvarg observed that "for people in desperate circumstances whose needs are not being met by the system, Wise Use has provided an identifiable enemy, 'the preservationist,' on which to focus their anger and vent their rage" (Helvarg 1994, 12). Although many environmentalists wrote off Wise Use as "astroturf," the movement possessed a real constituency. As one Wilderness Society report recognized,

> Environmentalists, who focus on wise use's conspiracy and vanguard messages and say the movement is too extreme to become a major force in America, overlook the power of the mainstream message. They also fail to see that the underside of the mainstream message paints a convincing portrait of environmentalists as the extremists in the debate over striking a balance between man and nature. (Switzer 1997, 193)

For several years in the early 1990s, Alliance for America and its allies organized a weeklong "Fly-In for Freedom," where Wise Use activists would converge on the nation's capital (Helvarg 1994, 123). The National Homebuilders Association "ran a full-page ad in national newspapers blaming the spotted owl for 'soaring lumber prices'" (Layzer 2016, 251). One *National Review* headline read, "Loggers on Welfare while Spotted Owls Flourish—But Ecologists Don't Give a Hoot" (Rice 1992). In a speech to the 1992 Republican National Convention (that, satirist Molly Ivins quipped, "would have sounded better in the original German"), Pat Buchanan bemoaned the efforts of "environmental extremists who put birds and rats and insects ahead of families, workers, and jobs" (Helvarg 1994, 39–40). The rightward momentum created by the Wise Use movement was particularly vital in the 1994 congressional elections, helping "Republicans pick up House seats in Arizona, California, Idaho, Nevada, Oregon, Utah and Washington, and to consolidate a Republican majority in the House and Senate" (Turner and Isenberg 2018, 79). The ultimate impact of Wise Use, as John Bellamy Foster has noted, was to provide "a new political base and 'populist' rationale for business-serving politicians seeking to undermine existing environmental laws" (2002, 106).

The ideological pull of this supposedly intractable struggle between "jobs and environmental protection" was indicative of a broader cooling of the relations between unions and environmentalists in the 1980s and 1990s. Similar struggles raged in the Southwest over environmental regulations on public lands, in Appalachia over the future of coal, and in the upper Midwest over the reauthorization of the Clean Air Act. Brian Obach writes that "job loss fears during the 1980s prevented unions from fully embracing environmental concerns, and the mainstream environmental movement adopted some conservative and defensive tendencies during this period as well" (2004, 53). While the movements came together in support of community right-to-know standards and worker health and safety, unions were divided over the desirability of the Clean Air Act amendments, drilling in Arctic National Wildlife Refuge, and Corporate Average Fuel Efficiency standards (Obach 2004, 59). But no issue loomed as large as climate change.

CLIMATE CHANGE AND CLASS STRUGGLE

The story of the conservative embrace of climate denial has been written many times over, but the key parts bear repeating. It didn't take long for capital—particularly fossil capital—to recognize that climate change was, for them, an existential crisis. Not only was business as usual threatened, but the very nature of the industry itself would have to change, or perhaps even cease existing. (Malm and the Zetkin Collective 2021). A 1998 Western Fuels Association report reflected on the intensity of the transformation that would be required if environmentalists and their allies were correct: "If one believes that increasing atmospheric concentrations of CO_2 place the world at risk, then the obvious solution is this: Society must de-carbonize, and the Fossil Fuel Age must end."

The resulting corporate mobilization spurred by the anxieties of fossil capital helped to provoke a sea change in the Republican Party. In spite of the antipathy of Reagan toward environmental policy writ large, his administration supported the Montreal Protocol to curb the hole in the ozone layer. During his campaign, George H. W. Bush pledged to reauthorize of the Clean Air Act and to "meet the Greenhouse Effect with the Whitehouse effect" (Balzar 1988). However, by the time James Hansen gave his famous 1988 testimony before the US Senate—popularly seen as the moment that climate change entered into mainstream political discourse—think tanks funded by fossil capital had been organizing in opposition to action on climate change for some time.[6] While the Bush administration ultimately supported the UN Framework Convention on Climate Change, which was ratified by the US Senate, Bush was subsequently praised by the business community for his efforts to keep "targets and timetables out of the convention" (Global Climate Coalition 1994).

In the first term of the Clinton administration, the nascent climate denial movement mobilized against a proposed BTU tax (on heat content of fuels, measured in British Thermal Units), included in a broader deficit reduction bill after West Virginia Democrat Robert Byrd had voiced strident opposition to a carbon tax. The National Association of Manufacturers immediately took aim at the measure, forming the American Energy Alliance to distill the message that the BTU tax was a business and job killer (Erlandson 1994, 179–180). Citizens for a Sound Economy—founded by fossil capitalists Charles and David Koch and funded by tobacco and petrochemical interests, but claiming to have 250,000 members—organized rallies and aired advertisements asserting that the bill would harm workers and "raise every family's tax bill" (Tuholske 1993; Vandewater 1993); utilities in Michigan and Nebraska ran full-page ads calling the tax a "job destroyer"; and "each Republican who rose to speak on the House floor declared the specific number of jobs that would be lost in his or her district due to the BTU tax" (Erlandson 1994, 179–180). The bill passed the House with the BTU tax intact, but the measure was ultimately pulled from the Senate bill after vulnerable Democrats balked. Some Democrats who lost in the landslide midterm elections (popularly termed the 1994 "Republican Revolution") attributed their defeat directly to their support for the BTU tax—a cautionary lesson that would be heeded in subsequent debates over cap and trade.

As the fight shifted explicitly to climate policy, the most influential of the early denial organizations was the Global Climate Coalition (GCC), organized by the National Association of Manufacturers and comprising a broad range of companies from the manufacturing and extractive sector, as well as several national trade associations, including the American Forest and Paper Association, the American Petroleum Institute, Chevron, Chrysler, DuPont, Edison Electric Institute, Exxon, Ford, GM, Shell Texaco, and the US Chamber of Commerce (Brulle 2022).[7] The GCC's stated mission was "to coordinate business participation in the international policy debate on the issue of global climate change and global warming" (GCC 2001). And coordinate it did, organizing a nationwide advertising blitz that sought to weaken support for the Kyoto Protocol. One 1997 GCC ad in the *New York Times* featured a group of smiling multiracial children who had a plea for Clinton: "Americans work hard for what we have, Mr. President—don't risk our economic future" (GCC 1997a).

Not satisfied to let the GCC work its twisted magic, many of the coalition's individual members also engaged in their own political maneuvering. For instance, at the very moment he was advocating for global free trade out of one side of his mouth, out of the other side, Chamber of Commerce President Thomas Donohue was expressing concern over declining national sovereignty: "America's jobs and economy are being held hostage to the environmental political agenda of the United Nations" (GCC 1997c). The carbon-based energy taxes needed to implement any US pledge would, according to the Global Climate Coalition, "provoke industry relocation, which would force jobs out of the United States" (GCC nd). In the *Washington Post*, an ad from the GCC-led Global Climate Information Project interjected that "the only thing this treaty cools down is America's economy" (GCC 1997b).

The general strategy of industry and conservative think tanks was one of job blackmail on steroids; if climate policy was passed—even a market-based policy—then the entire economy (not simply a particular company, sector, or region) would be damaged. According to the Heritage Foundation, if the United States were to meet its pledge under Kyoto (a 7 percent reduction in GHG emissions below 1990 levels), "20 to 30 percent of the basic chemical industry would move to developing countries, all primary aluminum smelters [would] close by 2010, there would be a 30 percent decline in steel producers, domestic paper production would be displaced by imports, [and] nearly a quarter of the cement industry would close" (Antonelli and Shafer 1997). In sum, "Americans [would] feel like they're living through the oil price shocks of the 1970s all over again" (Antonelli and Schaefer 1997, 16).

Many labor unions seized on the so-called developing country exemption—only the industrialized countries whose emissions were responsible for the bulk of the problem were obliged to make binding reductions—to build a case against Kyoto—a fact that capital took great pains to promote. The AFL-CIO opposed the protocol, while the United Mining Workers claimed that over one million jobs would be lost nationally (Global Climate Coalition 1998). In spite of the fact that the New Democrat Timothy Wirth, now serving as Clinton's undersecretary for global affairs, successfully lobbied to integrate market-based mechanisms into Kyoto, the Senate unanimously passed (95–0) the broad and ambiguous Byrd-Hagel Resolution (1997) asserting that the United States should not sign any climate agreement that does not also mandate cuts from the developing world or that "would result in serious harm to the economy of the United States." While the Clinton administration played a major role in negotiating the treaty, it was dead on arrival to the Hill.

The next decade-plus was one in which the United States backtracked—further deregulating energy markets under the oil-soaked reign of George W. Bush and Dick Cheney, while making only fleeting and half-hearted gestures toward climate policy. Amid the morass, cap and trade became the mechanism to which Democrats hitched their wagon; the path to passing climate policy by appealing to forward-thinkers in the business community and attempting to reach across the aisle to the moderate Republicans who remained. In introducing his 2003 cap-and-trade bill (cosponsored with John McCain), Joe Lieberman excitedly announced that "the Business Council for Sustainable Energy endorsed the concept that market-based climate policies can reduce gas emissions while promoting technology-based solutions, reduce energy dependence, and bolster the competitiveness of U.S. industry." While the market-based approach persuaded a handful of Republicans, even more Democrats voted against it. Five years later, a similar Lieberman-Warner bill not only failed to overcome a Republican filibuster, but also hemorrhaged moderate Democrat votes. In a letter to Senate Democratic leadership, ten Democratic senators echoed corporate talking points in suggesting that the bill hadn't done enough to "form partnership[s] with regulated industries to help them reduce emissions as they transition from an old energy economy to a new energy economy," and to provide relief for consumers in the form of "additional allowances to utilities."[8]

The election of Obama was, once again, cause for enormous hope among environmentalists that "this [would be] the moment when the rise of the oceans began to slow and our planet began to heal," as his campaign nomination speech put it. Such optimism was ultimately misplaced. The Democrats' 2009 cap-and-trade bill (Waxman-Markey), sought to apply the "lessons" of these past defeats by tacking further right. The most prominent coalition pushing for the bill was the U.S. Climate Action Partnership, comprising several dozen CEOs of major corporations (AIG, Alcoa, BP, Caterpillar, ConocoPhillips, Duke Energy, DuPont, Lehman Brothers, PG&E), as well as leaders of prominent centrist environmental organizations, like the Environmental Defense Fund, National Resources Defense Council, Nature Conservancy, and National Wildlife Federation. Supported by foundations, as well as hefty $100,000 annual fees from its members, the attempt to broker a "strange bedfellows" agreement between corporations and major environmental organizations was from the get-go "inherently asymmetrical"—the environmentalists gave and the corporations took (Skocpol 2013).

The bill was introduced in the midst of a brutal health-care fight that had unleashed yet another wave of corporate organizing, deepening party polarization and awakening something of a grassroots movement: the Tea Party. "While Obama's health-care bill was useful in riling up Tea Party protesters," wrote the journalist Jane Mayer, "his environmental and energy policies were the real target of many of the multi-millionaire and billionaires in the Koch circle" (2017, 245). Drawing on the lessons of the Wise Use movement, as well as their past experiences in the BTU and Kyoto fights, fossil-fuel front groups FreedomWorks and Americans for Prosperity worked to curate a grassroots presence, construct a working-class veneer, and a feign a middle American affect. Americans for Prosperity embarked on a nationwide Hot Air Tour against "cap and tax," wherein AfP president Tim Phillips and other assorted lunatics warned residents that "global warming alarmism" would lead to "lost jobs, higher taxes, and less freedom" (Americans for Prosperity nd). Parroting the long-standing rhetoric of the Competitive Enterprise Institute, California Republican Devin Nunes (2009) called it "the largest tax increase in American history" that would "destroy millions of jobs." As an alternative, he advocated additional oil and gas exploration that would create new, high-paying US jobs. Sensing a shift in the political winds of their home districts, Republicans in Congress, fearing primary challenges from the Tea Party flank, ran away from cap and trade like the plague.

Most major US unions, including the AFL-CIO, UAW, SEIU, Steelworkers, Laborers International Union of North America (LIUNA), and Communications Workers supported the cap-and-trade measure, while many corporations and trade associations, such as Murray Energy, the Chamber of Commerce, and the National Association of Manufacturers were opposed (Beutler 2009; Gemen 2009). Meanwhile, the corporations involved in the Climate Action Partnership were not all as supportive as they seemed. As Theda Skocpol points out,

> The corporations that participated in USCAP could double their bottom line bets—by participating in the strange-bedfellows effort to hammer out draft climate legislation that was as favorable as possible to their industry or their firms, and at the same time participating in business associations likely to lobby against much or all of the terms of that insider bargain once it faced Congress or the general public. (2013, 49)

With no major grassroots coalition working to pull the legislation to the Left, energize the environmental or Democratic Party base, and persuade ordinary people through actual on-the-ground organizing, it became a lopsided fight. The measure narrowly passed the House, but climate policy stalled—once again—in the Senate. The defeat of cap and trade left climate deniers giddy. As Myron Ebell, Competitive Enterprise Institute's director of energy and environment and chair of the Cooler Heads Coalition argued,

> There are whole doubts among the urban bicoastal elite but I think we've won the debate with the American people in the heartland, the people who get their hands dirty, people who dig up stuff, grow stuff, and make stuff for a living, . . . people who have a closer relationship to tangible reality. (Upin 2012)

This message is factually inaccurate—a strong majority of Americans, by this point, believed that climate change was occurring and was caused by the burning of fossil fuels—but it captures the general thrust of the Right's ideological strategy: frame environmentalists as elitists and workers as adamantly anti-environmentalist. Thanks to the minoritarian dimensions of US institutions—with, for example, single-member districts with first-past-the-post elections, and a Senate in which Wyoming has as many votes as California and a supermajority is required to take action—in order to pose a real obstacle to environmental progress, this alliance between extractive capital and the Republican Party need not persuade a majority of the US populace—only a sizable segment of the party's base.

CONCLUSION

From the late 1970s to the 2000-aughts, there existed broad bipartisan consensus with regard to core facets of the neoliberal agenda: markets were more efficient and cost-effective than state intervention; jobs and environmental protection could only coexist through the adoption of market-based forms of environmental governance; free trade was a win for workers and the environment; and the emergent "knowledge economy" was one in which economic growth could be decoupled from environmental impacts. Neoliberalism was only ever implemented selectively; it provided an ideological foundation that animated, but could never be squarely superimposed on, the deep historical grooves of US politics. The parties differed on the extent of their praise for markets and the facets of the state that they wanted to support. Democrats shifted focus to the meritocratic vision of the suburban set, while Republicans—in spite of a selective embrace of libertarian ideals (i.e., "society does not exist")—relied on an irrevocably communal imaginary: a nation made great by the middle-American workers of the heartland. As climate change intensified, this communal appeal shifted more explicitly to extractive workers—a nation indebted to "those who bring from the earth" (Republican National Committee 2016, 17). The message was clear: to make America great again, we needed to repay our debt to the coal miners, farmers, loggers, and roughnecks who have made our national flourishing possible.

But this narrative clashed with material reality in several vital respects. First, beneath the everyman veneer lay the ascendance of global financial capital, which sought not only to commodify everything in its wake but to turn every area of social and ecological life into a site of speculation. From free-trade agreements to cuts in capital gains taxes to financial services "modernization," both parties united to do the bidding of global finance. Second, beneath the superficial appearance of dematerialization lay the hard material realities—the devastating economic, environmental, and public health consequences—of a capitalist system tightly tethered to fossil fuels. The treadmill of production had gone global and sped up, producing profound hardships for the working class; not the selective view of the working class promoted in anti-environmental iconography but working class people across lines of race, gender, and nationality. The results were predictably tragic: unionization declined, wages stagnated, inequality skyrocketed, and environmental devastation reigned.

This attack on the working class occurred at a time when the treadmill of social reproduction was also being revamped and remade. If the prototypical treadmill of social reproduction lay in the suburban bedrooms of the early 1960s, by the 1980s, it had been sold off to a for-profit health club; more and more forms of social reproductive labor had been commodified. While we should take pains not to romanticize the sexism on which the mid-twentieth-century "family wage" was predicated—and while greater female participation in the workforce certainly represents important progress in the path toward gender equality—the broader structural parameters in which this progress occurred was one of capital's sharp pincers tightening its hold on working-class people: workers in the heavily unionized manufacturing sector were seeing their wages stagnate or being laid off, while a growing number of (disproportionately female) workers were being paid poor wages to do social reproductive labor (home health care, childcare, cleaning, etc.), before coming home to do more unpaid social reproductive labor (Dwyer 2013; Fraser 2017). These twin pressures—on production and social reproduction—resulted in a declining standard of living for the vast majority of Americans. The already astounding economic inequality in American life soared to stratospheric elevations. In 1980, the top decile made roughly 35 percent of total income in the United States; by 2010, they made nearly 50 percent (Piketty 2014, 291). In 1970, the top 1 percent earned less than 10 percent of the total national income; by 2000, this number had more than doubled (Piketty 2014, 300).

The one-sided class struggle of the era also produced enormous environmental injustices. The geopolitical turmoil of the late Cold War years left extractive interests looking inward: "In an irony that would escape few of the contemporary commentators, the seeming wastelands to which many Native Americans had been relocated or allowed to stay in the 19th century . . . were revealed by the Geological Survey to cover approximately half of the nation's uranium, one-third of its low-sulfur strippable coal, and a decent percent-age of its oil shale and natural gas" (Black 2018, 215; see also LaDuke 1999, 83–85). At the very same time the federal government was terminating its recognition of more than one hundred Native American tribes, the Bureau of Indian Affairs (BIA) was facilitating a new wave of mining in the west; one that powered urban centers and fueled corporate bottom lines, but left Native American communities struggling against the resulting pollution and toxic waste (LaDuke 1999; Gedicks 2001). Outside US borders, the Ogoni mobilized against contamination and theft by Shell in Nigeria, indigenous communities of Ecuador organized in the face of the toxic legacy of Exxon's operations and Rio Tinto/US oil and mining companies, and Indians reeled from a gas leak at a Union Carbide plant that killed thousands.

Meanwhile, the neighborhoods and communities of the poor and work-ing class were grappling with the historical aftermath of America's industrial golden years, a reality that groups like the Urban Environment Conference and Environmentalists for Full Employment had worked to call attention to. The period that I've covered in this chapter coincides with the rise of the envi-ronmental justice movement: in 1978, the toxic crisis at Love Canal galvanized national attention and helped to spur passage of the Superfund Act; in 1979 a uranium tailings operated by the United Nuclear Corporation in Church Rock, New Mexico, breached its dam, sending radioactive water downstream onto the Navajo Nation (Ivins 1979; Voyles 2015, 166–167); in 1982, resi-dents of Afton, North Carolina, were arrested protesting the state's plan to dump PCB-contaminated soil in this poor, predominantly African American locale; several years later, residents of Reveilletown, Louisiana—founded by freed slaves after the Civil War—organized Victims of a Toxic Environment United when an accident at a Georgia Gulf plant filled the air with plumes of vinyl chloride (Minchin 2003, 75); in West Dallas, African American com-munity members protested their continued exposure to pollution from a lead smelter (Bullard 1990, 45–51; 1993, 27). The list goes on.

While the history of movements that we can retrospectively recognize as engaged in struggles for environmental justice is long, the concepts of "envi-ronmental justice" and "environmental racism" were only now being intro-duced to activists, policymakers, and the general public. In 1979, the City

Care Conference was convened in Detroit by the National Urban League, Sierra Club, and Urban Environment Conference in order to strategize how to better form "coalitions to improve the urban environment" (City Care 1979). In 1983, the Urban Environmental Conference on the Toxic Threat to Minority Communities was held in New Orleans. In 1987, the United Church of Christ's Commission on Racial Justice released its seminal report, "Toxic Wastes and Race in the United States," finding that "people of color were far more likely to be exposed to commercial hazardous waste facilities and uncontrolled hazardous waste sites" (Lee 1993, 48; see also United Church of Christ 1987). In chapter 2, I argued that as the dominant justification for racial inequality shifted from biological to cultural difference, the natural wages of whiteness were institutionalized through white flight to suburbia, aided by suburban zoning laws that often integrated ostensibly environmental goals. In the late 1970s through early 2000s, even as upwardly mobile people of color moved to suburbia, entrenched patterns of residential segregation left a gap between the residential exposure to toxic pollutants experienced by the average African American, Hispanic, and Native American and that experienced by the average white person (see, e.g., Lester, Allen, and Hill 2001; Taylor 2014). As Robert Bullard long ago argued, "Housing discrimination, redlining, and other market forces [have made] it difficult for millions of households to buy their way out of polluted environments" (Bullard 1993, 21). This inequality was aggravated by the shift from the welfare state to the carceral state; the "ghetto environment" was locked into place—symbolically, through the rise of the underclass narrative, and materially, by the ascendance of the "prison fix" and tough on crime policies. The hollowed-out cores of inner cities would soon become sites of capital investment for predatory finance, but, for those left there, they were places of profound social and environmental injustice (Taylor 2019).

At the same time, however, when we disaggregate racial categories (i.e., white, Black, Hispanic, Asian, Native American, etc.) into their class dimensions (i.e., working-class white, working-class Black, working-class Hispanic, etc.), we start to see something of a shared vulnerability among working-class people of all races and nationalities emerge. This vulnerability was not equal— generally speaking, as a consequence of the history of settler colonialism and institutional racism that I've traced throughout the book, Native American, Hispanic, and Black workers were still more likely to be exposed to most pollutants than white workers—but exposure to unsafe levels of toxicity was, by this period, the norm for working-class people writ large. While the afore-mentioned United Church of Christ report shockingly found that "3 out of every 5 Black and Hispanic Americans lived in communities with uncontrolled toxic waste sites," less reported was the alarming conclusion that "*more than half of the total population of the United States reside[d] in communities with uncontrolled toxic waste sites*" (1987, xiv; emphasis added).

In a number of locales, this resulted in working-class struggles for envi-ronmental justice that built new coalitions across lines of race and gender. In *Forging a Common Bond*, historian Timothy Minchin (2003) details one such coalition. In 1984, when German chemical corporation BASF asked workers in its Geismar, Louisiana, plant to accept a broad range of concessions—a wage freeze, changes to seniority rights, paying a higher percentage of health-care costs—the OCAW local refused (NY Times editors 1986; Maraniss and Weisskopf 1987; Minchin 2003). The company responded by locking the workers out, alleging fears of industrial sabotage (Minchin 2003, chap. 2). The OCAW was a progressive union fresh off several campaigns that were successful thanks to the base of support its workers had managed to build among community groups and environmentalists. With that in mind, the union began to make a public case that the replacement workers brought in by BASF were untrained in maintaining the plant's health and safety and that the company had a track record of environmental violations at the Geismar plant and elsewhere. The union drew an explicit linkage between the terrain of production and social reproduction, the majority-white workers and the predominantly African American community members. As OCAW special projects manager Richard Leonard put it, "If the company is beating up its own employees, they're probably beating up everybody else too. . . . You make toxic chemicals, if they're going to take short-cuts, it's not only going to be with the workers, it's going to be with everybody else" (Minchin 2003, 60). Local community members formed Ascension Parish Residents Against Toxic Pollution, while the union worked to build broader coalitions with the Sierra Club, Greenpeace, the National Campaign Against Toxics, and the German

Greens. After five years, the workers gained a new contract, but perhaps just as important an outcome was the lasting bonds of solidarity that grew out of the engagement among workers, community members, and environmentalists. For example, the resulting Louisiana Labor-Neighbor Project (LLNP) would go on to successfully organize in a subsequent battle against Shintech Corporation's proposed PVC plant in St. James Parish (Minchin 2003, 163; see also Roberts and Toffolon-Weiss 2001).

These vital struggles, however, were largely outside the fray of US party politics. Rather than calling attention to the shared vulnerability of workers, the anti-environmental ideology of the day worked to exploit racial and gender difference in the interest of extractive capital and the bipartisan group of politicians doing its bidding. The New Democrats not only ignored the plight of the working class; their strategy of "triangulation" entailed an adoption of core elements of the conservative war on crime and war on workers, and a broader embrace of markets as the path toward social and ecological progress. Contrary to many analyses of the period, the historical struggle between the ruling class and the working class didn't disappear at this moment; it merely shifted forms as a rearguard revolt from capital responded to the threat of social movements that sought to embed markets within ecological limits and social protections via state regulation. As calls for strong state action in defense of the environment intensified, the defensive posturing of global fossil capital found ideological roots: in tradition, place, and images of a romanticized, "authentic" US worker. Building on the momentum of the Sagebrush Rebellion and Wise Use movement, the trope of the white male extractive laborer surged to the fore. As climate change loomed ever-larger, the calls to protect this selective vision of the US worker grew angrier and angrier.

4 UNDOING THE *OIKOS*, AWAKENING RESISTANCE: THE POLITICAL ENVIRONMENT OF "TRUMP COUNTRY"

In February 2020, a cap-and-trade bill passed out of the Oregon state legislature's budget committee and headed to the Senate floor for a vote. Senate Republicans, however, were nowhere to be found. Their absence was strategic—even though Democrats held a supermajority in both chambers of the state legislature, the Senate needed a quorum in order to undertake any business. And achieving a quorum required that at least a handful of Republicans be in attendance. This wasn't the first time that GOP senators had walked out in protest over proposed climate policy—the previous year they'd fled the state to prevent a vote on a similar cap-and-trade measure. When Democratic Governor Kate Brown sent state troopers to try to bring the negligent lawmakers back to their jobs, one Republican went so far as to threaten violence: "Send bachelors and come heavily armed," he warned (Zimmerman and Flaccus 2019).

While the majority of state's residents expressed displeasure at the GOP maneuvering, hundreds of supporters set up shop outside of the capitol building to show their support. Many held signs with messages for the legislators: "save our jobs," "our families matter," and "the only vaccine Oregon needs is one to prevent cap and trade." A pickup truck at the rally had "Long Live the Working-Class" emblazoned across its tailgate (KATU 2019; Profita 2020; Selsky 2020). Speaking in favor of the opposition, one rural county sheriff reminded Oregonians that "this state was built by the timber industry and by farms, ranchers, construction and other blue collar industries" (Dickinson 2019). The rally was organized by the organization TimberUnity, which touts its working-class support from truckers, farmers, and loggers but is also funded by the oil and gas and timber industries and has close ties with Three-Percenters, Proud Boys, and QAnon, as well as those who participated in the occupation of the Malheur Wildlife Refuge (Sunshine 2020).

These linkages with the Far Right didn't stop higher-ups in the Trump administration from holding several prominent meetings with the group (KATU 2021). In return, TimberUnity has demonstrated its fealty to Trump—at least one board member attended the January 6 insurrection at the US Capitol, and many others have continued to voice their support for his false claim that the 2020 election was stolen (Leber and Breland 2020). As environmental journalist David Roberts astutely observed, the Republican Party, Far Right groups, and resource industries have become "an interconnected network in the state" (Roberts 2020; see also Davis 2019; Kormann 2019; Benham 2020; VanderHart 2020). Unfortunately, these interconnections extend far beyond Oregon.

Although such linkages point toward troubling shifts in the Republican Party, with ominous impacts on the future of US democracy, the voices of groups like TimberUnity do not reflect the views of the vast majority of working-class people in the country. In fact, even in many conservative-leaning rural and suburban towns and cities across the United States—the oft-invoked "Trump country"—residents are clamoring for environmental protections and rallying against extractive interests. In 2018, in the small town of Stephenson, Michigan, opponents of a proposed open-pit sulfide mine packed a high school gymnasium to address their concerns to the state Department of Environmental Quality (Engel 2018). In rural Virginia, a coalition of locals have, for years now, engaged in tree-sits and protests in efforts to stop the 300-mile Mountain Valley Pipeline from being constructed (Earth First! 2018). At a 2018 EPA "community engagement" event that I attended, resident after resident of Rust Belt manufacturing towns—like Hoosick Falls, New York, and Merrimack, New Hampshire—stepped up to the microphone to forcefully condemn corporate environmental malfeasance, as well as the agency's insufficient response to drinking water contaminated by per- and poly-fluoroalkyl substances (PFAS) (Therrien 2018).

These examples of environmental activism are indicative of a broader trend: as environmental degradation has increased in scope and intensity, resistance is emerging in working-class locales that have otherwise served as strongholds of conservatism and emblems of anti-environmentalism. And, far from reflecting the unpopular ideas of a loud minority, there is reason to believe that—at least in some cases—the activists' views represent widespread public sentiment in the area.[1] How might we understand the paradox of intense and widespread environmental activism emanating from communities that staunchly supported Trump, a candidate who brazenly embraced anti-environmentalism? And what might these untimely outbursts have to teach us about the relationship between the working class and the future of US democracy?

I begin attempting to answer these questions by engaging with the sprawling literature, both academic and popular, that seeks to understand the rise of Trumpism and the decline of US democracy. My analysis pivots between academic texts and popular commentators who have zoomed in on so-called Trump country and its "white working-class" inhabitants—like those mobilized to protest the Oregon cap-and-trade bill—as the key to unraveling the reactionary moment in which we find ourselves. I argue that although the dominant portrayal of Trump country captures real grievances that are being voiced by residents of rural, conservative-leaning areas, it is deficient in numerous regards: (1) tending to focus on the politics of right-wing activists (disproportionately white men employed in extractive industries or heavy manufacturing) rather than the majority of residents; (2) failing to capture the demographic and ideological diversity that exists within the locality or industry being depicted; (3) rarely taking note of the social forces that are actively attempting to thwart the Right's extractive visions; and (4) neglecting to consider the ample support for the right-wing politics emanating from gated communities and upper-middle-class suburban Chambers of Commerce. To ameliorate these aporias, I shift focus to those who are engaged in environmental resistance against the anti-environmental agenda of Trumpism. Drawing on interviews with environmental activists from conservative locales, as well as insights from environmental justice and eco-feminist theory,

I suggest that, far from being irrevocably wedded to a reactionary politics, many residents of "Trump country" are outraged at the deterioration of their homes and local communities—what I term, "the undoing of the *oikos*"—but they remain politically ambivalent, alienated, and malleable. If this is correct, then a renewed Left focus on grassroots environmental politics could awaken political transformation in people and places that many liberals have been quick to write off as irredeemable, potentially reinvigorating working-class consciousness in the process.

NEOLIBERALISM AND DEMOCRACY: THE ROAD TO TRUMPDOM

In 1944, amid the blasts and blood of global war, two influential books appeared, both of which purported to explain the rise of fascism and its relationship to the economic and geopolitical ruptures of the interwar period. The first was Friedrich Hayek's *The Road to Serfdom*, which made a straightforward case that the steady incursion of the state into the "free and spontaneous" operation of markets created the conditions of possibility for fascism. "Few are ready to recognize," Hayek mused, "that the rise of fascism and naziism was not a reaction against the social trends of the preceding period but a necessary outcome of those tendencies" (1944, 6). For Hayek, the idea of society itself was a dangerous invention—the mythical collective actor in whose name the great sins of our age had been committed. In the interest of society—the common good, the general welfare, the public interest—economic activity becomes directed toward collective goals (like a guaranteed minimum income, universal health care, and full employment) that trample on the rights and interests of some and require the coercive force of the state to implement. The end result, according to Hayek, is the progressive encroachment of the state into more and more areas of life, and the continual diminution of individual liberty.

The second book was Karl Polanyi's *The Great Transformation*, which made the opposite argument: markets had become the principle around which modern life was organized, and the result was an "avalanche of social dislocation" that produced several varieties of social blowback, including fascism (1944,

42). Whereas preindustrial societies refused to leave land, labor, and money to the whims of supply and demand (or at least protected these areas of life from some of the most pernicious effects of commodification), over the course of the nineteenth and early twentieth centuries, these "fictitious commodities" were increasingly left to the invisible hand of the market, with devastating impacts on both ordinary people and on nature. Unlike Hayek, Polanyi viewed free markets not as the path toward freedom but as utopian abstractions that had never, and *could never*, truly exist; when the quest for "improvement" through markets invariably produces threats to "habitation"—that is, ordinary people's ability to live dignified lives—society takes "measure[s] to protect itself" (1944, 3). Polanyi referred to this as a "double movement" or a "counter-movement"; markets not only require state intervention, but the dominance of markets inevitably produces resistance from society.

While Hayek's neoliberal screed was serialized in the *Reader's Digest*, attention to Polanyi's democratic socialist tome was largely confined to the academy. This changed over the course of the 1980s and 1990s as the institutionalization of Hayek's prescriptions—first touted by the scions of capital in the late 1930s and then propelled forward by the conservative think tanks of the 1970s and 1980s—wreaked havoc on the lives of ordinary people. Amid the hollowing out of industrial cities, the decimation of unions, the skyrocketing of inequality, and the ominous warnings of a climate catastrophe to come, many commentators turned to Polanyi to suggest that neoliberalism had undone the regulatory state that was required to keep corporations in check and the welfare state that was needed to shield vulnerable populations from the "creative destruction" of the market. The market had become disembedded from social relations, with brutal social and environmental costs. Over the past several decades, scholars and activists alike have pointed toward stirrings of resistance—anti-globalization protests, mobilizations for racial justice, gender equality, indigenous and immigrants' rights, and Occupy Wall Street—as evidence of an ascendant Polanyian countermovement that would soon contest neoliberalism's dominance. But, amid the continued decline of unions, the simultaneous repression and corporate capture of movements for women's rights, racial justice, and environmental protection, and the resurgence of right-wing parties around the world, some began to question whether neoliberalism had inflicted so much damage on society that it had now become difficult to envision a way out.

The most rigorous and sophisticated of such analyses has come from political theorist Wendy Brown, who asserts that after three-plus decades of neoliberal dominance, the *demos*—that is, *the people* who possess the capacity for self-rule—is coming apart at the seams. Building on the work of French philosopher Michel Foucault, the central thrust of Brown's argument is that the neoliberal mode of governance creates political subjects (that is, those individuals living under its domain) unable to think their way out of a neoliberal world—who are either incapable of engaging in responsible dialogue over the common good or who perceive the common good through the metric of the market (Brown 2015). Insofar as both governmental and societal institutions have been reconfigured by neoliberalism, the political subjects who are socialized within their purview are themselves put through something of a structural adjustment program. The subjects interpellated by neoliberal institutions are "entrepreneurs of the self" laboring to increase their human capital in a society structured by competition (Foucault 2008). As Foucault observes, "The individual's life itself—with his relationships to his private property, for example, with his family, household, insurance and retirement—must make him into a sort of permanent and multiple enterprise" (Foucault 2008, 241). Brown argues that, under neoliberalism,

> both persons and states are construed on the model of the contemporary firm, both persons and states are expected to comport themselves in ways that maximize their capital value in the present and enhance their future value, and both persons and states do so through practices of entrepreneurialism, self-investment, and/or attracting investors. (2015, 22)

Central to this enterprise form is the concept of *human capital*, the knowledge, education, and training that impact one's wage-earning potential, and its entrance into the rationalities of the state and citizen alike. The role of the state, from a neoliberal perspective, is no longer to mediate relations between labor and capital, but to increase human capital: "Thus all the problems of health care and public hygiene must, or at any rate, can be rethought as elements which may or may not improve human capital" (Foucault 2008, 230). This emphasis on human capital also extends into the family. Chicago school economist Gary Becker made the case that, like firms, the behavior of families—the number of children they choose to have, who enters the workforce, the amount of attention and type of education provided to children, and so on—is guided by an economic rationality; "irrational" behavior can therefore best be corrected through the incentives provided by the market (Becker 1993). Melinda Cooper contends that, for Becker,

> the family in its equilibrium state can be understood as serving a kind of natural insurance function that is disturbed when the welfare state socializes insurance. . . . The postwar welfare state [has destroyed] the natural altruism of the family, . . . [but] the decline in welfare initiated by Reagan will ultimately compel the poor to restore the bonds of kinship as a source of privatized welfare. (2017, 60)

Foucault claimed that the goal of the neoliberals is "to bring labor into the field of economic analysis . . . to put [themselves] in the position of the person who works" (2008, 223); as Cooper points out, this applies to both productive and social reproductive labor. Contra a Marxist perspective, this does not mean considering how labor power is sold and surplus value is wrested away by the owners of the means of production; rather, neoliberalism views income as a return on one's own capital investment. This is not a conception of labor power but of *capital-ability*; if flourishing personally and professionally depends on one's skill set, then, over time, "the worker himself appears as a sort of enterprise for himself" (Foucault 2008, 225). The political effect is an erasure of labor as a social concept: "When everything is capital, labor disappears as a category, as does its collective form, class, taking with it the analytic basis for alienation, exploitation, and association among laborers" (Brown 2015, 38).

According to Wendy Brown, Foucault highlighted the shifts from a classical liberal to a neoliberal subject (the gradual expansion and intensification of *homo oeconomicus*), but he was unable to foresee the centrality of finance capital to contemporary political economic relations. "Today, homo oeconomicus . . . has been significantly reshaped as *financialized human capital*: its project is to self-invest in ways that enhance its value or to attract investors through constant attention to its actual or figurative credit rating, and to do this across every sphere of existence" (Brown 2015, 33). The neoliberal subject, economic historian Phillip Mirowski concurs, "is a jumble of assets to be invested, nurtured, managed, and developed; but equally an offsetting inventory of liabilities to be pruned, outsourced, shorted, hedged against, and minimized" (2013, 108). Brown believes that liberalism has long been a struggle between *homo oeconomicus* (man as an economic actor) and *homo politicus* (man as a collective participant in the polis). But for the first four centuries it was a fair fight: "The prominence of man's economic features in modern thought and practice reconfigures without extinguishing his political features—again, these include deliberation, belonging, aspirational sovereignty, concern with the common and with one's relation to justice in the common" (Brown 2015, 94). Today, however, it appears that *homo economicus* has won, once and for all: "Neoliberal reason, ubiquitous today in statecraft and the workplace, in jurisprudence, education, culture, and a vast range of quotidian activity, is converting the distinctly *political* character, meaning, and operation of democracy's constituent elements into *economic* ones" (2015, 17). In such a context, Brown asks, "how do subjects reduced to human capital reach for or even wish for popular power?" The answer, perhaps, is through support for demagogues driven by violent and racist forms of nationalism, yearning for the purported golden age of "traditional" gender and sexual norms, and willing to accomplish such ends through antidemocratic means.

Is Trump's election proof that *homo oeconomicus* has triumphed and the demos has come undone? Brown asserts that "neoliberalism generates a condition of politics absent democratic institutions that would support a democratic public and all that such a public represents at its best: informed passion, respectful deliberation, aspirational sovereignty, sharp containment of powers that would overrule or undermine it" (2015, 39). It is difficult to argue that our current political order comprises any of these vital elements of a democratic public; rather, uninformed passion, angry discord, and anxious apathy appear to reign. The decimation of society under neoliberalism laid the foundation for a Trump victory and, more broadly, the global rise of the Far Right. As Cornel West (2016) put it, "The neoliberal era in the United States ended with a neofascist bang." George Monbiot (2016) contends that "the backlash against neoliberalism's crushing of political choice has elevated just the kind of man that Hayek worshipped." Brown (2018) asserts that

> the ground for Trump's rise was tilled not just by neoliberalism's destruction of viable lives and futures for working and middle-class populations through the global outsourcing of jobs, the race to the bottom in wages and taxes, and the destruction of public goods, including education. This ground was also tilled by neoliberalism's valorization of markets and morals and its devaluation of democracy and politics, Constitutionalism and social justice.

In the years since we first witnessed the rise of "Trumpism," an entire genre of scholarship has emerged to make sense of the causal forces behind his victory and of the people and places that make up his base of supporters. There is also a clear environmental dimension to the concept of "Trump country" that warrants a brief discussion.

THE NATURE OF "TRUMP COUNTRY"

On March 13, 2016, with the campaign for presidency in full swing, Hillary Clinton, stood before a town hall in Columbus, Ohio, and made the following remark:

> I'm the only candidate [who] has a policy about how to bring economic opportunity using clean renewable energy as the key into coal country. Because we're gonna put a lot of coal miners and coal companies out of business. . . . And we're gonna make it clear that we don't want to forget those people. Those people labored in those mines for generations, losing their health, often losing their lives, to turn on our lights and power our factories. . . . Now we've gotta move away from coal and all those fossil fuels, but I don't wanna move away from the people who did their best to produce the energy that we relied on. (CNN 2016)

Selectively seizing on one facet of this pledge, the media reported Clinton's "attack on coal country" with aplomb. Conservatives responded with predictable outrage. *The Federalist* dutifully passed along Clinton's "message for Coal Miners: You're Fired" (Payton 2016). The *NY Post* mused that "Hillary's vow to kill coal miners' jobs finishes a vast Democratic betrayal."

Calling her comments "callous" and "wrong," Mitch McConnell suggested that they underlined "the need to stand up for hard working, middle-class coal families" (Bruggers and Gerth 2016).

For the remainder of the campaign, Trump delighted in repeating her supposed gaffe, juxtaposing it with his own unabashed praise for coal miners. The site of his first speech after wrapping up the Republican Party nomination for president was Charleston, West Virginia. Flanked by coal workers wearing hard hats and union patches, surrounded by "Trump Digs Coal" and "The Silent Majority Stands with Trump" campaign signs, he opened by proclaiming his intention to "put the miners back to work" (C-SPAN 2016). Reports of Trump's rally—and images of its attendees—circulated widely in both liberal and conservative media outlets. While the commentaries differed in their tone and, in some cases, critique, the storylines shared striking similarities: the "white working class" was the agent behind Trumpism; America First the rallying cry; economic and racial anxiety the causal impetus; and blue-collar jobs the palliative (Colvin 2016; Falders and Smith 2016; Jacobs 2016).

That these were extractive jobs, and that America First here excluded putting America first in environmental protection garnered surprisingly little attention.[2] What did attract enormous coverage were the white male workers of "Trump country." Throughout the campaign and in the aftermath of the election, journalists reporting from "Trump country"—Main Street in Logan County, West Virginia; Main Street in McDowell County, West Virginia; a diner in Pikeville, Kentucky; a bar in Clarksburg, West Virginia; a barbershop in Buchanan County, Virginia; and so on—continually profiled the rise of reactionary politics among the residents of Appalachian coal country and, to a lesser extent, the Rust Belt and the rural west (Kaplan 2016; Lewis, Silverstone, and Sambamurthy 2016; MacFarquhar 2016; McLelland 2016; Saward 2016; Beckett 2017; Von Drehle 2018). The places of Trump country, according to one *New York Times* article, "have little in common but economic hardship, a sense of longing for the better times they once had and an unshakable belief that a President Trump might be the answer to their troubles" (Kaplan 2016). And yet, the people interviewed in the aforementioned profiles of Trump country—a plaster and stucco contractor, a former textile factory worker forced into early retirement, a boot factory worker, and many small business owners, miners, loggers, and construction workers—had quite a bit in common: they generally worked in blue-collar jobs, were overwhelmingly (though not exclusively) white, and disproportionately men.

Even before his victory in the general election, an old demographic descriptor had resurfaced to both explain and indict the reactionary and self-defeating politics of this demographic: the white working class (WWC) (see, e.g., Cohn 2016; Waldman 2016; Rothenberg 2019).[3] Residing in old industrial strongholds, like Scranton and Wilkes-Barre, Pennsylvania, and Youngstown, Ohio, as well as Appalachian coal country, the WWC was most often conceptualized in terms of education level (referring to whites without college degrees) and less frequently in terms of income or occupation. The dominant framing, in both conservative and liberal outlets, was of a WWC facing economic hardship, the deterioration of communal ties, and the plague of addiction. The WWC perceive themselves to be falling behind, both economically and culturally. They feel "'uneasy and out of place' in their own country" (Davis and Ballhaus 2016); believe that the bonds holding their communities together are dissolving (Macfarquhar 2016); and think that they've become "strangers in their own land" (Hochschild 2016). Sensing that their home is no longer what it once was, they articulate a close connection between the deterioration of local neighborhoods and communities and that of the US nation. Alienated by decades of war and economic decline, they glom on to Trump's overt white supremacy, seeking to Make America Great Again through the exclusion of immigrants, and the return of manufacturing and extractive sector jobs (McCammon 2015; Macfarquhar 2016).[4] Trump, as critic turned apologist J. D. Vance put it, "is the candidate of the man who opens his morning paper to find that another of his neighbors has died of a heroin overdose, . . . of the proud coal miner who voted for Bill Clinton and then watched as his wife promised to 'put a lot of coal miners and coal companies out of business'" (Vance 2016).

Accompanying the near ubiquity of depictions of coal miners holding Trump banners and oil workers in MAGA hats, several notable analyses of Trumpism also picked up on a deep-seated anti-environmentalism that resides within the Trump voter (Hochschild 2016; see also Cramer 2016, 155–158; Gest 2016, 83–84).[5] The WWC residents of Trump country, according to a

Wall Street Journal article, "blame the Obama administration's regulations for a downturn in the coal industry and figure Mr. Trump will quash those rules and ignore scientists who warn about global warming" (Davis and Ballhaus 2016). Accompanying and intensifying the opposition to environmental regulations is a long-standing and powerful romanticization of particular types of extractive labor that I described in detail in chapter 3. In her analysis of Trump voters in Wisconsin, for example, Katherine Cramer finds that her interviewees' have a narrow definition of what constitutes socially valuable labor:

> People are like: Are you sitting behind a desk all day? Well that's not hard work. Hard work is someone like me—I'm a logger, I get up at 4:30 and break my back. For my entire life that's what I'm doing. (Guo 2016)

The anti-environmentalism of Trump country is particularly evident in Arlie Hochschild's celebrated ethnography of Louisiana Tea Partiers, *Strangers in Their Own Land* (2016). Many of Hochschild's interviewees are threatened by water contamination due to the area's close proximity to the oil and gas and chemical industry. They love their bayou environment for both its aesthetic beauty and material sustenance; it is integral to their livelihood and leisure and is a part of their deeply rooted attachment to place. The conservative state of Louisiana has responded to the widespread pollution by emphasizing end-of-pipe fixes, like a pamphlet on "how to trim, grill and eat mercury-soaked fish" (Hochschild ibid., 110–111). The residents find themselves saddened by what has occurred but nonetheless voting for Tea Partiers who want to abolish the EPA (see also Malin 2014, Jerolmack and Walker 2018).

Such findings suggest that, in decimating society, neoliberalism has also changed the way that the white working-class residents of Trump country relate to nature itself: jobs and environmental protection appear as a zero sum game ("Bring Back Coal!"); nature is reduced to a pool of resources to be commodified and put to use ("Drill Baby, Drill!"); environmental regulations are seen to harm nature and society alike by removing incentives provided by private property; regulatory agencies are viewed as bloated and self-serving; environmentalists are in cahoots with the urban elites against the Real America; and the acceptance of environmental risks is the assumed cost of living in the modern world. As one Hochschild interviewee, unhappy about the pollution but unwilling to support strengthened environmental regulations, put it, "Pollution is the sacrifice we make for capitalism" (2016, 179). From this perspective, the hegemony of neoliberalism seems near total, and the hope for the future of both democracy and a habitable planet appears bleak.

What, then, ought to be done to combat the politics of Trumpism? "What kinds of Left political critique and vision," asks Wendy Brown, "might reach and transform" those who support such candidates and projects? (2019, 188). The sheer intensity of neoliberalism's assaults on society—the rise of conspiratorial political currents, like QAnon, the power of dark money, the entrenchment of institutional barriers to democracy, the apparent intensity of red/blue and rural/urban divides—has left many in the academy searching for an answer to this question in the places they know best (abstract theory) or in the populations that reflect the worst (Far Right movements, the Tea Party, etc.). The problem is not that their answers are wrong (though, as I detail below, they often are), but that their solutions are predicated on an incomplete picture of US political life.

The reality is that the attitudes and behaviors of the working-class residents of conservative-leaning areas of the country are significantly more complex and nuanced than these popular tropes would have it. On the one hand, in the 2016 election, Trump voters were far more likely to live in or near the town in which they were born (Cox and Jones 2016). And while the median Trump voter was comfortably middle class in economic terms (Carnes and Lupu 2017), they were more likely to live in areas that were downwardly mobile (Chinni 2017). There is a correlation between counties with high opiate use and support for Trump (Goodwin et al. 2016), as well as counties where life expectancy has declined in recent years and support for Trump (Bor 2017). On the other hand, however, "at least since the 1980s, white working-class Americans have never made up a majority of Republican voters in presidential elections. . . . Lower-income white voters without college degrees aren't a majority of Republican voters, and they aren't increasing as a share of GOP voters" (Lupu and Carnes 2021). "The majority of Trump's voters," reminds anthropologist Jessica Smith (2017), "were actually college-educated, middle- and upper-class whites." Most of the poorest and most environmentally ravaged communities in the country didn't vote for Trump (Silver 2016). And, even among those in which a majority of voters did, "the myth that Trump voters are inscrutable and monolithic" (Catte 2019) veils a much more complex and interesting reality. Further, among a portion of the communities that did vote for Trump, the extraction of natural resources and pollution of water and air—whether already occurring or looming—are provoking forms of resistance that are neither hyper-nationalist nor neoliberal.

Within the existing scholarly and popular literatures, you'd be hard pressed to know this. Unreflexively parroting conservative political imaginaries, many on the liberal and progressive Left have presented a narrative of sprawling conservative spaces (the heartland, middle America, Trump country, etc.) nearly devoid of political struggle, when, in fact, the reactionary forces of so-called Trump country are being actively resisted by many. We should be careful not to overstate the pull of the organizations and movements that are pushing against the grain in places now dominated by conservative politics, but a failure to recognize and learn from these unlikely insurgencies incapacitates the very critiques and political visions that we leftists so desperately seek. How might we orient our theoretical lenses in a way that allows us to glimpse the mere possibility of progressive change? Building on one of Polanyi's profound insights—that when social and environmental protections are ravaged by markets, at least some segment of society rises up in opposition—we might refocus on the resistance being spurred by threats to habitation.

WHERE THE WORKING CLASS CALLS HOME: THE *OIKOS*

The words "economy" (*oikonomia*) and "ecology" (*oekologie*) share a common Greek root: *oikos*. For the ancient Greeks, the *oikos* referred to the household, which existed in sharp contrast to the *polis*, where humans committed to "living together in a deliberately governed fashion, to self-rule in a settled association that comprises yet exceeds basic needs, and to the location of human freedom and human perfectibility in political life" (Brown 2015, 87). The *oikos* was the space of necessity, ruled by the patriarch and filled with "natural" inequalities and power asymmetries; the *polis* was the space of appearance, where free debate among equals over the proper way of living together and achieving "the good life" reigned (Arendt 1998 [1958], 36–37). The *polis* was ontologically prior to the *oikos*—in the sense that humanity itself was defined by its capacity for political life—but participation in the *polis* required that one was first freed from the necessities of the *oikos*. The Greek household comprised "relations of rule and relations of production" (Brown 2015, 88) and *oikonomia* simply referred to the proper husbanding of material resources for the household (Mitchell 2008). "Status in the *polis*," Habermas writes, was "based upon status as the unlimited master of an *oikos* (*oikodespotes*)" ([1962] 1989, 3).

This naturalization of enormous social inequality—and concomitant relegation of labor relations to a place outside of politics—is but one of myriad flaws of the ancient Greeks, as well as many of their modern interlocuters, such as Arendt. Economic affairs were nonetheless limited by Aristotelian morality: "Wealth is never to become its own end" and "wealth that is accumulated for its own sake is unnatural" (Brown 2015, 89). Trading was acceptable only insofar as it was necessary for a household's self-sufficiency and the maintenance of a community (Polanyi 1944, 56–57; see also Jowett 1885, xxiv–xxv). Polanyi observed that Aristotle offered a critique of "incipient market trading at its very first appearance in the history of civilization" (1957, 67). As classical liberalism was institutionalized in the seventeenth and eighteenth centuries, the *oikos/polis* dichotomy was transposed onto the private/public divide—with all the exclusionary baggage that entailed. The public sphere, where citizens discuss and debate pressing issues, was constructed as the locus of civilization, rationality, masculinity, and progress, while the private, or "domestic," sphere was taken to be governed by the laws of nature, emotion, femininity, and tradition. It is at this moment that economic activities, which had previously been confined to the space of necessity, were "permitted to appear in public" (Arendt 1998, 46) and political economy—"the knowledge and practice required for governing the state and managing its population and resources"—entered into existence (Mitchell 2008, 1116). As Habermas notes,

> The term "economic" itself, which until the seventeenth century was limited to the sphere of tasks proper to the *oikodespotes*, . . . now, in the context of a practice of running a business in accord with principles of profitability, took on its modern meaning. . . . Modern economics was no longer oriented to the *oikos*; the market had replaced the household, and it became commercial economics. (1989, 20)

It is no mere coincidence that the term *ecology* emerged in the mid-nineteenth century, as the industrial revolution shook the foundations not only of the global organization of production but the reproduction of great swathes of life itself, human and nonhuman.[6] The maturation of industrial capitalism over the course of the nineteenth century unchained the economy from both the *oikos* and the "excessive" interventions of the state—a reality enabled by the ideological naturalization of the economy as an autonomous sphere of life, and the reduction of nature to a resource to be used in pursuit of human progress. As Jason Moore has argued, the two uses of *oikos*—the economic and the ecological—have, at least in dominant forms of liberal thought, been treated as separate realms of life and objects of analysis, the former hedged firmly within "the social" and the latter the sole purview of "the natural." A core innovation of socialist political economy, however, was to suggest that social labor and the natural environment are metabolically linked—the dialectical relationship between the two forms the basis of individual and collective life itself. Their commodification, driven by the imperatives of capital accumulation, is thus the site of intense and ongoing political struggle (see, e.g., Polanyi 1944; Marx and Engels 1992; Fraser 2017). A great deal of the theoretical and popular commentary that I've reviewed in this chapter has lost sight of this fact, by either emptying the *oikos* of its political content or reducing it to a site of reactionary fervor. In doing so, such work has rendered growing waves of discontent and resistance virtually unintelligible. Politicizing the *oikos*—and making sense of this resistance—requires that we undertake several tasks: (1) examining the interdependent relationship between human homes (the social) and nonhuman habitats (the ecological); (2) tracing the linkages between the labor that occurs at home (the site social reproduction) and at the workplace (the site of production); and (3) considering how we might build bonds of solidarity between movements structured around threats to the home and neighborhood and those structured around conditions of work. To provide conceptual tools for this work of politicization, I turn to eco-feminism, environmental justice scholarship, and socialist feminism.

The Home as a Socio-Ecological Site

Human homes and nonhuman habitats have always been entangled in myriad ways. Our homes comprise our relationships with our family, roommates, and neighbors; wood, cement, and wires; electric cables that link us to power grids; water lines that connect our communities to springs, lakes, and reservoirs; the grass and trees in our yards and neighborhoods; and the myriad nonhuman species (birds, rabbits, squirrels, etc.) that pass through these yards and neighborhoods or share them with us. This interconnection is something that environmental justice and eco-feminist scholarship has long grasped. Eco-feminist Mary Mellor observes that

> it is interesting that while Haeckel chose a name based on the Greek *oikos*, meaning household or dwelling, it was [late nineteenth-century American scientist] Ellen Swallow who showed the direct connection between daily domestic life and the environment. . . . For Swallow, the importance of educating women was that the home, even more than the workplace, was where primary resources such as nutrition, water, sewage, and air could be monitored. (1997, 14)

In environmental struggles, which—at a grassroots level—have always been disproportionately led by women, entrance into the public sphere via collective action frequently occurs because of threats to the home. Rachel Stein observes that "because environmental ills strike *home* for vulnerable communities, and because women have often been responsible for that domain, women engage in these movements in order to protect and restore the well-being of families and communities threatened by environmental hazards or deprived of natural resources needed to sustain life and culture" (2004, 2). Robert Gottlieb similarly details how environmental justice activists have often referred to themselves as "home-makers" in their efforts to achieve clean air and water, healthy food, and access to environmental amenities (2005, 275). Insofar as homemaking is a practice and a process, comprising "securing food, clothing, and shelter, caring cooking, cleaning, making, provisioning, nurturing, teaching, welcoming, excluding, fighting, living, and dying" (Meyer 2015, 143), the home is a powerful site in which our lived realities collide with political ideologies and projects sculpted from on high. As John Meyer argues, "Our material experiences as home dwellers are an important subject of citizenship in general and of citizen action to promote sustainability in particular" (2015, 158). In reflecting on her path to activism after the Three Mile Island nuclear accident, labor-environmental activist Jane Perkins put it bluntly: "When I was forced to leave home, I was quickly politicized. I became involved" (Hernandez 1981).

There is ample evidence that the past forty years of neoliberalism has wreaked havoc on our homes; the social and ecological relations that constitute the *oikos* are coming apart at the seams. For many residents of the United States, our homes—and the neighborhoods and communities in which they are embedded—are threatened by violence, addiction, gentrification, toxic contamination, and/or climate change. The data suggesting that we are experiencing the undoing of the *oikos* is overwhelming: over the past several years, the United States has averaged almost two mass shootings a day and rates of violent crime have crept up, threatening to undo decades of progress (*New York Times* 2023); nearly a third of all US households now spend more than 30 percent of their income on housing (Whitney 2023); after seeing notable declines for over a decade, rates of homelessness have risen dramatically since 2017 (National Alliance to End Homelessness 2023); from 1990 to 2020, the cost of childcare increased 220 percent, now averaging over $10,000/year for a single child (ChildCare Aware 2021); deaths from opioid overdoses rose from fewer than ten thousand in 2000 to more than eighty thousand in 2021 (National Institute on Drug Abuse 2023); the drinking water of the majority of Americans is contaminated with unsafe levels of PFAS (Andrews and Naidenko 2020); and so on. This is to say nothing of climate change. A ProPublica study suggests that, within the next fifty years, half of Americans will experience a decline in their environment due to climate change and "for 93 million of them, the changes could be particularly severe" (Lustgarten 2020). Generally speaking, it will be hotter, and there will be more prolonged droughts, and more outbreaks of severe weather. Some parts of the country will almost certainly become uninhabitable because of sea level rise (along the coasts) or extreme temperatures (in the Southwest in particular).

In short, working-class Americans are experiencing threats to habitation that—while certainly differing in many respects from those experienced by English peasants during the enclosure movement—are astounding in their speed and intensity. But has this contemporary "avalanche of social dislocation" produced a Polanyian countermovement? In many contemporary renderings of US politics, the *oikos* has been transformed into a space of vice, or one of insular communalism. In terms of the former, the myriad exposés of opiate addiction in the Rust Belt provide glimpses into an *oikos* where people binge drink, do drugs, and play video games in attempts to escape their gloomy lives (see, e.g., Trent and Robertson 2018; Achenbach 2019); in terms of the latter, profiles from "Trump country" depict an *oikos* where underemployed white workers rage against immigrants while clinging to fraying communal ties and praying for extractive jobs (Lewis, Silverstone, and Sambamurthy 2016; MacFarquhar 2016).

There is ample evidence that the undoing of the *oikos* has generated political activism; the direction of this activism, however, is an open question. It is abundantly clear that a politics constructed around protection of the home can be depoliticizing or exclusionary. For example, one can conceive of the *oikos* as a "political" site through a neoliberal lens and attempt to consume one's way out of its destruction. To ameliorate environmental threats, we take shorter showers, eat less meat, and ride our bikes to work. To ward off unhappiness, addiction, or loneliness, we exercise more and meditate, or perhaps buy ourselves new entertainment systems. To deal with rising costs of living, we work both more efficiently and longer. These personal acts are not necessarily negative (in fact some are undoubtedly good for the self and planet); the point is that they fail to lead us into collective action, and, on their own, only reinforce the political economy that we should be aiming to resist.

Of even greater concern, the anxieties surrounding the home have often been conjured into exclusionary forms of political community. The rise of eco-fascism reminds us that climate-induced threats to the homeland (*heimat*) are rapidly being translated into a language of "blood and soil" (Malm and the Zetkin Collective 2021). More frequently, however, the anxieties emanating from the *oikos* work alongside more banal (and anthropocentric) commitments to nationalism that aim to shore up supposedly declining wages of whiteness by doubling down on dreams of fossil-fueled flourishing amid industrial transformation and demographic shifts. As Brown (2019) puts it, the partisans of Trumpism "cling to the soil, even if it is planted in suburban lawn devastated by droughts and floods from global warming, littered with the paraphernalia of addictive painkillers, and adjacent to crumbling schools, abandoned factories, terminal futures" (2019, 187). Such a characterization aptly captures the spirit of Trumpism, but conversations with environmental activists who work in so-called Trump country have given me reason to believe that something more interesting is also occurring.

Environmental Revolt from "Trump Country"

"I've grown up around coal mines almost my entire life," Dustin White told me, recounting the generations of his family who worked in the mines. A lifelong resident of West Virginia, he supported coal from an early age; "we always heard the 'coal keeps the lights on' narrative," he explained, citing the pro-coal lesson plans that he was besieged with from elementary school on up. This changed abruptly one day when he ran into his uncle, who told him about a mountain-top removal site on nearby Cook Mountain:

> I grew up on the foot of Cook Mountain in a holler. I knew very much about that mountain. We spent a lot of time up there hunting, morel mushroom hunting, recreating on ATVs, and there's a couple cemeteries up there that are 200 years old. Those graves include both my 8th and 7th great grandfather . . . [the latter] a union soldier in the Civil War. My uncle proceeded to tell me that the coal company had blocked off the access road to the cemetery and were saying that there were no cemeteries up there. That was the first aha moment. (White 2021)

Dustin soon linked up with a local environmental group who arranged for him to take a fly-over of the site:

> We took off from Charleston and when we crossed over the Boone County line, you could just start to see these massive expanses of just moonscapes that were once mountaintops. . . . We finally got over to Cook Mountain and we could see this little island of trees in this vast rock and debris field where my ancestors lay buried. That really upset me, and I got back on the ground and started doing more research. (White 2021)

He began connecting the devastation of coal mining to the health problems that were so prevalent in the community in which had grown up, due to coal slurry that had contaminated the ground water. Since that moment—for over a decade now—he has been on the front lines of environmental struggles in West Virginia.

Dustin's story is exceptional, but the broad contours of it fit a pattern found in the lives of many environmental justice activists. A person becomes aware of environmental contamination that impacts their home, kin, and community. This threat to a place to which they have intense emotional ties motivates them to take action. Previously apolitical or even conservative in their politics, they get involved in a local struggle. As they participate in collective dialogues and actions, they forge connections with other activists and start to build bonds of solidarity. Their politics almost invariably shift Left, but—in a community in which jobs are scarce and conservative political values are dominant—their political interventions are carefully curated to appeal to the material needs of the local population rather than rigid adherence to any party platform or political ideology. They take seriously the need for decent-paying jobs and understand the ways in which particular industries have become woven into local identities. Among these activists, there is often a deep understanding of the need to create linkages between the community-based struggles of environmentalists and the workplace struggles of labor unions.

Veronica Coptis, for example, grew up in a working-class family who moved from Pittsburgh to a tiny town in western *Greene* County, Pennsylvania, when she was young. In rural Pennsylvania, she lived alongside a waste valley fill from a coal mine:

> I only knew about it because I was learning how to drive in that valley in backroads . . . and all of sudden we weren't allowed to go down into that valley anymore. . . . The locals were kind of like, "that's what happens." . . . And I was really angry. All the wildlife that I'd seen [while] recreating in those valleys, all of a sudden, their habitats were gone. (Coptis 2021)

Shaken by this experience, she went to college to study environmental science, but all that she learned about organizing, she told me, she learned from waiting tables. And as she was working her shift at the diner one day, one of her regulars wanted to find out more information about a fish kill that had occurred nearby: "Muskies the size of shovels were showing up dead, and it kept expanding upstream," she explained. She helped her customer do some research, and they stumbled upon an organization called the Center for Coalfield Justice, which she now runs.

Ashley Funk's story has similarities to those of Veronica and Dustin. "I grew up poor—I wouldn't even be considered as working class," she explained to me. "I grew up with an ash dump, an old coal refuse dump in our back yard."

> The one house I lived in 'till I was around 7 or 8 years old. The ash dump, people would ride their four wheelers and dirt-bikes and quads all over it. . . . The dust would blow up into the air and in our house. . . . Our one neighbor was dying of lung cancer, there were a lot of concerns about how this was impacting people's health. . . . There are pictures of me as a kid just covered in black dust, just smiling, because we would play in it all the time. (Funk 2021)

These early experiences didn't take long to influence her thinking; as a teenager, Ashley joined a lawsuit against the state of Pennsylvania for failing to act on climate change. Today she directs the Mountain Watershed Alliance, an environmental organization in southwest Pennsylvania that seeks to protect and preserve the Youghiogheny River watersheds.

For Karan Ireland, environmental activism began with the rashes she and her children were experiencing. In 2014, she recalled, "we had a drinking water contamination [in Charleston, West Virginia] . . . a chemical leak into our drinking water source that affected 300,000 people. I got sick from taking a shower, my kids had rashes and headaches, and I was a single mom to two fairly young school-age children." Dissatisfied with the reaction of the Democratic governor—who took pains to minimize the relationship between the leak of MCHM (a chemical used to wash coal) and the coal industry—Karan took action.

> I started to post a lot on social media, . . . went to the Town Hall and was live Facebooking and tweeting what was happening. My friend and I decided to have a community meeting. We didn't know what we were doing but it was just like, "hey are you mad like we are mad, and what can we do about it?" We thought we'd have like five or six people come, but people heard about it. Ultimately, it blew up. . . . It got away from us. We ended up with like 150 people. That was really the beginning of something that I've never really stopped. (Ireland 2021)

She was eventually elected to the Charleston city council, and then became a campaign representative with the Sierra Club Beyond Coal and Beyond Dirty Fuels campaigns.

How might we understand the pathways that shook these activists out of their private lives and into lives of environmental activism? Reflecting on Marx's statement that "revolutions are the locomotives of world history," Walter Benjamin once mused, "perhaps revolutions are not the train ride, but the human race grabbing for the emergency break" (Benjamin 2003, 402). The *oikos*, we might say, is the emergency brake on what Jason Moore terms capitalist world-ecological history. The home—the place where we reside, whether we own, rent, camp, or squat—is one that we have deep attachments to; our most intense and meaningful relationships are often formed and enacted close to home. Environmental activism, in its various iterations, has often been a response to threats facing the home. From the Diné and Hopi struggling to save their homes from strip mining to suburban residents organizing in opposition to the bulldozing of open space and from Love Canal and Warren County to Standing Rock and Flint, the undoing of the *oikos* has long given rise to resistance. To be clear, I do not mean to draw a direct equivalence between the struggles waged by suburbanites clinging to middle-class status who find their backyard gardens suddenly awash in PFAS and, say, the longtime residents of "cancer alley" or members of the Navajo Nation dealing with lead in their water. The weight of history presses down on particular populations in uneven ways that need to be acknowledged in our collective struggles (lest we re-embed patterns of inequality in our contemporary strategic interventions and in our visions for the future). But it is equally myopic to focus exclusively on these differences while ignoring the fact that the status quo works for none of the aforementioned groups. The widespread material deprivation that characterizes today's world is where the potential for movement-building on a mass scale lies.

Take, for example, opposition to the Back Forty Mine, a proposed open-pit sulfide mine along the Menominee River of upper Michigan and northeast Wisconsin. Although the vast majority of potentially affected areas voted resoundingly for Trump in both 2016 and 2020, the mine is wildly unpopular among locals, and has engendered frequent protests, demonstrations, and the formation of coalitions between Native Americans from the Menominee Nation and the predominantly white, working-class communities nearby (Burie 2018; Anahkwet 2023; Gedicks 2023). Opposition to the mine is grounded in a desire to protect the home and the community in which it is irrevocably enmeshed. For members of the Menominee nation, the Menominee River is not only a source of drinking water and recreation, but also a vital part of their common home; the mouth of the river is their historical origin place. As Menominee activist Anahkwet (2023) explained to me, the river is also the site of "twenty-four of our cultural sites and three known burial sites of our ancestors."[7]

The stories of the non-Native community activists differ in important respects—lacking the same historical grounding and communal referent—but they nearly always start at home too: one family bought a retirement home on the river, another farms nearby, another has a family cottage on the banks of the Menominee, and so on. What is interesting is that this initial concern for a relatively private home-space is quickly conjoined to concern over the human and nonhuman inhabitants of nearby areas, including those of the Menominee Nation (Burie 2018, Engel 2018). Far from an individualistic or an exclusionary conception of home, the campaign to "save the Menominee" is functioning to build new social bonds, and even pushing some white residents to grapple with the area's settler colonial legacy (Gedicks 2018).

In rural parts of upstate New York and southwest Vermont, recent years have witnessed similar stirrings of resistance in working-class communities faced with toxic contamination of their water, land, and bodies. Anthropologist David Bond recounts how after the water sources of many residents of Hoosick Falls, New York, and Bennington, Vermont, were found to be contaminated with PFOA, residents fought back against the corporations responsible and the state agencies who'd failed to hold them accountable. These mobilizations began at home, but it wasn't long before they jumped scales—from town halls to national environmental conferences, and from local collaborations to transnational coalitions:

> Over the past three years, the largely white, working-class communities of Hoosick Falls and Bennington have hosted mothers from Flint, Michigan; sent care

packages to the water protectors at Standing Rock; collaborated with high schoolers from East LA working on drinking-water issues; published op-eds in communities around the US discovering PFOA in their water; and reached out to communities around similar plastics plants in India and China. (Bond 2021, 13)

Similar forms of activism have sprung up across the country. The Keystone XL pipeline was eventually stopped not merely because of Joe Biden's executive order, but because of years of protests and pressure on the part of an unlikely coalition of ranchers, farmers, indigenous activists, and climate change advocates (Bosworth 2022). "In Washington and Oregon," observes geographer Zoltan Grossman, "Native nations are using their treaty rights to stop plans to build coal and oil export terminals. . . . The same largely white fishing groups in that region that used to aggressively protest treaty rights now back the tribes in protecting fisheries from oil and coal shipping, and in restoring fish habitat damaged by development" (Grossman 2019; see also Grossman 2017). In Minnesota, Wisconsin, and Upper Michigan, indigenous citizen-activists have teamed up with environmentalists and local landowners to try to stop the Enbridge Line 3 and Line 5 pipelines (Engelfried 2020; Stopline3.org nd). Even in rural Pennsylvania, where residents once welcomed oil and gas companies with open arms, support for fracking has been waning as the impacts of natural gas extraction on their land, water, and homes have become increasingly apparent (Jerolmack 2021).

The aforementioned examples suggest that the forms of resistance emanating from the *oikos* have the potential to lead to broader political transformation. Political theorist and activist Romand Coles highlights the role of the home within grassroots movements opposed to neoliberalism. The meetings of concerned community members often take place in the homes of activists themselves, before eventually moving into larger venues like schools, workplaces, public libraries, and community centers: "There are one-on-one meetings in which listening to and sharing profound narratives about our sources of inspiration and aspiration begins to generate powerful new connections, senses, and possibilities" (Coles 2016, 170).

Such an everyday politics—starting at home and then moving out into the community—sometimes even leads to the explosions of untimeliness, of *shock politics* (mass protests, general strikes, etc.), that we retrospectively view as important historical events. Indeed, it is through the "call and response" between the everyday politics of the *oikos* and the evanescent politics of the mass demonstration that the potential for transformation is heightened. Insofar as threats to the *oikos* politicize ordinary people—turning homemakers into community-builders—this is a politics "teeming with natality" (Coles 2016, 174), which can lead to the reconfiguration of the *polis* and the revitalization of the demos. Out of the *oikos*, "[a] thawing demos is stimulated and begins to move" (Coles 2016, 170).

Environmental and Labor Movements, Unite!

In order to bend this movement toward an emancipatory politics, though, today's social movements cannot neglect the realm of production. Matthew Huber (2019) points out that environmental justice struggles have been waged for decades now with only limited success, at least in part because the movement writ large has neglected "the strategic question of how to translate local livelihood concerns into a broader mass environmental movement able to take on capital." At the same time, however, at a moment where rates of unionization in the United States hover around 10 percent (Bureau of Labor Statistics 2023), successful workplace organization requires cross-movement coalition-building and broad public support. In this vein, activists must intervene to highlight the material connections between the home and the workplace, the labor of social reproduction and production. Silvia Federici suggests that cultivating macropolitical transformation "requires first a profound transformation in our everyday life, in order to recombine what the social division of labor in capitalism has separated" (2012, 144). Noting the historical relationships between organized labor and urban communities of color, Robin D. G. Kelley observes that "working people spend countless hours in non-work settings such as neighborhoods, homes, in transit, in urban public spaces, in houses of worship, in bars, clubs, barbershops, hair salons, in various retail outlets, not to mention, welfare offices, courtrooms, even jail cells" (1999, 43). Kelley reminds us that the most effective workplace struggles—the famous sit-down strikes in Akron and Flint, struggles of Latina and Asian-Pacific cannery workers in the west, strikes among Black washerwomen in the South—were utterly dependent on grassroots community institutions: "Mutual benefit organizations, fraternal organizations, and religious groups not only helped families with basic survival needs, but created and sustained bonds of fellowship, mutual support networks, and a collectivist ethos that ultimately informed working-class political struggle" (1999, 43–44). As I detailed in chapter 3, drawing on the case of the OCAW and community members of Ascension Parish, the division between the home and the workplace is something that many of the most successful labor *and* environmental justice movements have worked to bridge by drawing on widespread opposition to the state and corporate actors endangering both, and working to cultivate a shared commitment to clean air and water and a safe and dignified workplace.[8]

This strategy has the potential to effectively combat the continued power of the "jobs versus environment" narrative that has been so effectively deployed by the Right. "Most of the time," Ashley Funk explains, "people go back to the jobs narrative."

> And it's a narrative that drives me bonkers because it's not true. When people say that rural communities like ours need jobs, what they mean is they need well-paying jobs. And that only is the case because so much labor that currently exists is undervalued. In all these communities, no matter what, you need a grocery store, you need a hairdresser, you need people who are gonna take care of the elderly, gonna take care of kids. . . . Most of the time you have teachers, plumbers, electricians. When you think about a community and all of the jobs that you need to sustain that, . . . those jobs exist, but for some reason we have as a society said that we think it's okay to pay a regular laborer $7.25 an hour if they work at a restaurant or service job. . . . People shouldn't have to risk their lives and their bodies and their community's well-being and their health in order to have a family-sustaining job. (Funk 2021)[9]

In this regard, Funk's insight meshes with that of socialist feminists who problematize the devaluation of labor that has been historically gendered as women's work. While wreaking havoc on all workers, neoliberalism has dramatically intensified the pressure put on social reproductive labor. Financialized capitalism, Nancy Fraser explains, has externalized the care work on which capital accumulation depends "onto families and communities while . . . diminish[ing] their capacity to perform it" (2017, 32). From such a perspective, it is no surprise that the commodified sectors of social reproductive labor—for example, the service industry, teachers, nurses, and flight attendants—are also hotbeds of contemporary labor activism and among the most progressive unions on climate policy.[10] So called "pink-collar labor" not only provides a material bridge for connecting the *oikos* to the workplace, it offers a pathway to a more sustainable political economic order. As Alyssa Battistoni argues,

> The kind of work that we'll need more of in a climate-stable future is work that's oriented toward sustaining and improving human life as well as the lives of other species who share our world. That means teaching, gardening, cooking, and nursing: work that makes people's lives better without consuming vast amounts of resources, generating significant carbon emissions, or producing huge amounts of stuff. (2017a)

Attention to what Battistoni (2017b) calls *oikos work*—the undervalued labor done by those engaged in social reproduction and by nonhuman lives and forces—challenges the continued prevalence of a working-class iconography that emphasizes white male workers in resource extraction and heavy manufacturing (without writing off such workers as unworthy of our compassion and solidarity). It also underscores the need for a more ethical relationship with the nonhuman lives that have historically been reduced to the material substratum propelling human advancement. This work of building a more just, sustainable world necessarily happens within the realms of production and social reproduction. Rather than reifying an artificial boundary that is itself the product of capitalism, eco-socialist movements must work to bring these separate spheres of conflict together into one mutual struggle.

The most transformative calls for a Green New Deal recognize this, emphasizing the need for work-life balance; options for well-compensated, meaningful work; the expansion of labor rights; and the provision of care for working-class communities and the environment. A far cry from the cap-and-trade proposals that I discussed in chapter 3, the 2019 Ocasio-Cortez-Markey resolution for a Green New Deal (GND) tethers its call for "millions of good, high-wage jobs" to an effort to guarantee that everyone can enjoy "clean air and water, climate and community resiliency, healthy food, access to nature, and a sustainable environment." In this vein, the economist Paulina Tcherneva has proposed a federal job guarantee in the form of a "National Care Act," which underscores that we'll need to create green jobs not only in infrastructure, but also in ecological and social restoration. The act would include a twenty-first-century Civilian Conservation Corps, where the un- and underemployed would be given jobs working to prevent soil erosion and control flooding through ecosystem restoration, to monitor species, work on CSA farms and community gardens, weatherize homes, compost, clean up vacant properties, restore historical sites, and provide childcare, elderly care, and after-school programs (Tcherneva 2018, 17–19; see also Aronoff et al. 2019, 65–77). These policies must provide *just compensation*—in the form of wages, benefits, and working conditions—so that, as productivity increases, we can all work less, turning down the two treadmills to a sustainable pace, rather than having to crank them up to the point that our minds and bodies break (Aronoff et al. 2019, 71–72). Finally, as Nick Estes (2019) points out, indigenous peoples have long engaged in vital care work for the environment and, at a moment where resource extraction is ramping up in the name of renewable energy, any GND worth its salt will need to institutionalize protection of indigenous sovereignty and prior-informed consent. While several of my interviewees noted that support for a Green New Deal in the communities that I've discussed in this chapter is undercut by the project's partisan connotations (Coptis 2021; Overton 2021; Stetson 2021), many of the concrete policies lodged under the banner of a GND continue to enjoy strong support.

CONCLUSION: WORKING-CLASS ENVIRONMENTALISM AND US DEMOCRACY

The progressive efforts to rethink and rebuild the *oikos* that I've outlined above face organized opposition. American conservatives recognize the power of home and have anchored their politics in denunciations of the looming threats facing the "traditional" *oikos* (like nontraditional gender roles, neighborhood and school integration, immigration, and, increasingly, environmentalism).[11] For the political Right, the response to the undoing of the *oikos* is to connect extant threats to local places to the broader threats supposedly imperiling the traditional nation: globalism, cosmopolitanism, multiculturalism, "transhumanism," critical race theory, and eco-socialism. As the conservative journalist-cum-white supremacist Sam Francis wrote, years ago, "Economics . . . derives from Greek words meaning 'household management,' and the purpose of economic life . . . is not simply to gain material satisfaction but to support families and the social institutions and identities that evolve from families as the fundamental units of society and human action." In a 1996 essay titled "From Household to Nation," Francis made the case for a politics that would connect the traditional social morality that emanated from the middle-American household to an economic nationalism that resonated with ordinary people—a project that began to take hold under Trump and finds its clearest expression in the work of contemporary National Conservatives and post-liberals who have railed "against the dead consensus" that privileges individual autonomy over family stability and communal solidarity (Various Authors, 2019). A 2019 broadside against "consensus conservatism"—signed by prominent conservative intellectuals like Sohrab Ahmari, Rod Dreher, and Patrick Deneen—concludes by stating, simply, "We believe home matters."

As environmental crises intensify, it is not surprising that (anti)environmental politics would begin to enter more centrally into the Right's political calculations; thus far, in the form of economic nationalism tethered to fossil capitalism. In this regard, Polanyi was prescient. The *American Prospect's* Robert Kuttner points out that, while Marx expected the contradictions of capitalism to result in a proletarian revolution, "Polanyi with nearly a century more history to draw on, appreciated that the greater likelihood was fascism" (Kuttner 2017). In response to the upheavals produced by the dominance of markets, segments of society would seek to re-embed the market within an aggressive and aggrieved form of nationalism. Today, finding their homes decimated by neoliberalism, many have sought solace in the backward-looking sense of home provided by fossil nationalism. We see such a politics in coal miners rallying for Trump, in local community members suffering from toxic air and water who insist that pollution is the price we pay for freedom, and in groups like TimberUnity, reflective of the growing linkages between fossil capital and Far Right organizations more generally (Nobel 2020).

Polanyi was acutely alert to the potential for reactionary movements to safeguard the conditions for habitation by adopting insular identities and seeking to actualize violent fantasies, but he was no fatalist; the possibility for socialism remained. By contrast, accounts of Trumpism and many recent analyses of a US democracy imperiled by the forces of neoliberalism and/or neo-fascism tend to rest on oversimplified tropes of acquiescent demographics incapable of mounting resistance or preternaturally wedded to their anti-environmental identities as "white working class" or, even worse, "white working-class men"— essentialized subject positions whose supposed fossilization obfuscates the ruling-class fuel that drives the Right's anti-environmental commitments. For a growing number of commentators on the broadly construed Left, society is so irreparably damaged that it is difficult to imagine how its defense could take anything other than authoritarian forms. Instead, we're stuck searching for emancipation via ontological fixes (e.g., discursive and aesthetic interventions that transgress the binary between nature and culture) glommed onto romanticized and essentialized visions of subaltern subject positions (Bessire and Bond 2014; Johnson 2017, 2019). The resulting strategic prescriptions are long on theoretical posturing and short on concrete political interventions.

This chapter has tried to sketch out the beginnings of an alternative theoretical orientation grounded in the leftist tradition of immanent critique. Our analyses of possible "ways out" of our current conjuncture might productively begin with the glaring contradictions lodged within the material realities of working-class populations that sit at the interstices of the political struggles of our day: ordinary people, across lines of race, nationality, and gender, whose daily experiences with the realities of economic precarity and environmental uncertainty—and whose frustration with the inability of either political party to provide viable solutions—has left them ambivalent when it comes to party politics, but also angry and ready for change. If this is so, then there are crucial lessons to be learned from environmental activists whose work unfolds in conservative-dominated places. I am not suggesting that the predominantly white, rural activists interviewed in this chapter should be seen as constituting the vanguard of leftist activism; rather, my argument is that they offer important insight into leftist strategy as we grasp for ways to build broad-based mass movements and majoritarian coalitions.

In highlighting cases of environmental organizing and resistance within so-called Trump country, there is a risk of painting too rosy a picture. Wendy Brown's analysis of our current conjuncture is right on many counts; as the spread of neoliberalism "evacuates the content from liberal democracy and transforms the meaning of democracy *tout court*, it subdues democratic desires and imperils democratic dreams" (2015, 44). Decades of tax cuts, austerity, deregulation, and privatization have created numerous points of blockage in efforts to cultivate a flourishing, just, and sustainable democracy: transforming education into job training; decimating organized labor and making workplace struggles more difficult to wage; stripping the state of its capacity to check corporate power; and convincing citizens that pollution is the price we pay for consumer sovereignty and the fossil-fueled freedoms that are enjoyed by many. There exists no shortage of places across the United States where outbursts of environmental activism like those outlined above have failed to materialize in the wake of mega-extractive projects, where anti-environmental voices rule the day, or where environmental resistance continues to face a steep uphill battle.

The environmental mobilizations that do occur in conservative-dominated, rural areas are oftentimes unsuccessful. In spite of the herculean efforts of activists to forge broad-based, indigenous-led movements to stop oil and gas infrastructure from being constructed, both the Dakota Access and Enbridge Line 3 pipelines are now operational. When environmental and community activists used environmental laws to mount successful legal challenges to the Mountain Valley Pipeline, NextEra Energy Resources determined, in 2022, that "the continued legal and regulatory challenges have resulted in a very low probability of pipeline completion" (Kirong 2022). Not long after, however, West Virginia Senator Joe Manchin finagled promises of pipeline permitting reform into the Inflation Reduction Act, and in May 2023, the debt-ceiling deal between Biden and congressional Democrats effectively green-lighted the project.

And yet, there have been victories as well. In July 2020, when Dominion and Duke Energy canceled plans for the Atlantic Coast Pipeline, the Trump-appointed secretary of energy gave the credit to activists, claiming that "the well-funded, obstructionist environmental lobby has successfully killed the Atlantic Coast Pipeline" (Duffy and Levitt 2020). In June 2021,

Canadian energy infrastructure company TC Energy abandoned plans for the Keystone XL pipeline after the Biden administration revoked a permit for construction that had been granted by the Trump administration (White House 2021; Lindwall and Denchak 2022). And as of June 2023, resistance to the Back Forty mine has been successful. The environmental permits that had been granted to the owners of the mine were revoked by a Michigan state judge after seven months of evidentiary hearings. With the prospects of a new protracted, permitting process looming, the owners then sold the mine (EarthJustice 2021; Schulte 2021; Ellison 2022). Reflecting on several decades of resistance to the project, longtime environmental and indigenous rights activist Al Gedicks concluded our interview by expressing optimism for the future. The new owners, he explained, "still are not anywhere near submitting applications for a new permit process. . . . The company that had the longest track record in going through the permit process is bankrupt."[12]

Moreover, examples like the Dakota Access, Enbridge Line 3, and Mountain Valley Pipelines make clear that the most profound barriers to rejuvenating US democracy and saving the environment are institutional rather than ideological. These extractive projects all entered into existence via a mix of bureaucratic and judicial rulings and corrupt political giveaways that short-circuited democratic practice. There is more resistance to the forces of anti-environmentalism than perhaps ever before in this country. And this resistance is not monopolized by young urban activists, suburbanites aghast at right-wing transgressions against decorum, or liberal commentators of the professional managerial class. The undoing of the *oikos* is steadily forging a shared set of material interests among the growing ranks of the working class, across lines of race, gender, and nationality, who lack the resources to distance themselves from contemporary environmental threats. The adherence of working-class whites to the racial and class interests constitutive of fossil capitalism has widened inequality and diminished opportunities for living the good life in an increasingly warming and toxic world. The moment is ripe for a leftist intervention around an alternative vision of a flourishing *oikos*: where high-quality housing, childcare, education, health care, and access to clean air and water are universal rights; where working-class people can democratically organize productive and social reproductive labor; and where living beings are valued not on the basis of occupation or ascribed identity, but on their multifaceted contributions to the socio-ecological community.

Contra popular depictions, many residents of the communities that I've discussed in this chapter are not efficient and entrepreneurial neoliberals, apolitical and antisocial automatons, or obedient proto-fascists. The *oikos*—the socio-ecological place that we call home—interrupts the neoliberal ideology-subjectivity feedback loop that too many scholars and commentators see as almost hermetically sealed off from political intervention. Home is where the deepest socio-ecological connective tissue resides, frustrating neoliberal efforts to individualize and instrumentalize it. Indeed, the undoing of the *oikos* is already leading to promising political action. The challenge for the Left today is to capitalize on these fractured yet multiplying moments of "fugitive democracy" (Wolin 1996) and bring them together in the service of political transformation. In the ruins of the *oikos*, a reinvigorated and radicalized *demos* could emerge.

CONCLUSION: FROM ASHES TO FASCISTS?

On August 16, 2022, US President Joe Biden signed the Inflation Reduction Act (IRA) into law, arguably the first substantive piece of climate policy that the United States has passed in the four decades plus that we've been debating the issue. The centerpiece of the policy is nearly $400 billion dedicated to climate change mitigation, largely through the mechanism of tax credits on green technology and energy efficiency improvements for individuals and corporations alike. Not only was the policy tethered to the interests of middle-class homeowners and private fortunes, but—to the dismay of environmentalists—it included a requirement to expedite oil and gas leases, as well as pledges of Democratic support for one of fossil capital's biggest pet projects: oil and gas pipeline permitting "reform" (Derman 2022; Handler and Bazilian 2022). At its core, the IRA is fraught with contradiction—a cause for both relief and consternation. It is, as several commentators noted, a potential "game changer" that could provide the support for renewables needed to propel the United States down the path toward decarbonization. It is also, as a number of progressive environmental organizations pointed out, riddled with fossil fuel giveaways that could well thwart the Biden administration's quasi-ambitious climate and environmental justice pledges (Rees 2022). Even if the IRA does provoke an energy transition, it is unlikely to be a just one.

How did we reach a point where a centrist policy that pales in comparison to the crisis before us seems like a cause for celebration? This question is a microcosm of the puzzle that I've grappled with over the course of the book: why, in the face of rampant socio-ecological injustice—the theft of nature, the displacement of environmental burdens onto working-class people and their communities, and the ongoing destruction of ecological systems—has fossil capital been so successful in its continued quest to create a world in its own image? This concluding chapter begins by shedding light on contemporary climate politics in the United States, before reaching back into history and summarizing some core insights into the nature of US class ecological relations that I've sketched out over the course of the book.

FROM THE GREEN NEW DEAL TO THE INFLATION REDUCTION ACT

In February 2019, Representative Alexandria Ocasio-Cortez and Senator Ed Markey introduced a nonbinding resolution for a Green New Deal (GND)—a vision for an ambitious Democratic Socialist initiative that would marry climate policy to a jobs program and policies aiming to achieve greater social equity, such as universal health care, jobs for all, and funding for communities beset by environmental injustice. Over the next year, this vision coalesced into concrete policy proposals, such as that laid out by Bernie Sanders' 2020 presidential campaign. The Sanders GND called for a $16.3-trillion-dollar public investment over ten years that would wed climate policy to concrete support for front line communities and a just transition for fossil fuel workers (Sanders 2020).

While Sanders eventually lost the nomination, Biden integrated aspects of GND proposals into his Build Back Better agenda. After his election, Build Back Better was eventually parceled out into three parts: (1) a COVID-relief and stimulus package (the American Rescue Plan); (2) an infrastructure bill; and (3) a social and climate policy bill. After the first of these passed, with only modest environmental provisions, infrastructure and social and climate policy shifted to the center of debate. While progressives pushed for a simultaneous vote on the infrastructure and social and climate policy bills, moderates insisted on prioritizing the former, with the explicit promise of an eventual vote on the latter. Biden acceded to this request, and the Infrastructure Investment and Jobs Act was signed into law in November 2021. The House almost immediately took up—and eventually passed—a scaled-down social climate and policy bill (rebranded the Build Back Better Act) along party lines. The bill, however, faced opposition in the Senate not only from a united Republican front but from centrist Democrats Joe Manchin and Kyrsten Sinema. Sinema was ultimately lured into support (after Democratic leaders jettisoned plans to close the carried interest loophole that allows financial firms to pay less taxes than most workers) (Schwartz 2022b), but on December 19, Manchin—whose electoral and personal fortunes are deeply tied to fossil fuels—announced on right-wing Fox News that he was "a no" (Finn and Tsirkin 2021). The measure seemed all but dead until a mid-summer announcement by New York Senator Chuck Schumer; a grand bargain with Manchin had been struck (Everett and Levine 2022). In order to avoid a filibuster, the new bill, strategically reframed as the Inflation Reduction Act, passed the Senate through the budget reconciliation process along strictly partisan lines. All Democrats in both chambers of Congress voted for it; all Republicans voted against it. Biden signed the bill on August 16, 2022, remarking that it was "one of the most significant laws in our history." After enshrining this landmark climate policy into law, he immediately turned and handed his pen to Manchin. The man who had in recent election cycles received more money from fossil fuel companies than any other member of Congress—who, a decade earlier, had filmed a campaign advertisement in which he held up a rifle and literally fired a bullet through the cap-and-trade bill under debate—had now become the face of US climate policy (Manchin 2010; Milman 2021).

In contrast to their vehement opposition to the Green New Deal (DiChristopher 2019), the scions of fossil capital were surprisingly bullish on the IRA. The CEO of Occidental Petroleum touted the act as "one of the most transformative bills ever signed" (El Dahan and Strupczewski 2023). On an earnings call, Chevron executives "predicted continued growth in the Gulf and tied that directly to being able 'to lease and acquire additional acreage'" (Brown and Phillis 2022). As an article in Forbes pointed out, the IRA takes a "more now, less tomorrow" approach to oil and gas, which "should allow domestic oil and gas companies to transition at a reasonable pace, providing space for continued investment in oil and natural gas supply while encouraging the industry to advance decarbonization plans over time." The measure also subsidizes fossil capital's own preferred path toward decarbonization "by pumping money into nascent technologies like carbon capture and storage (CCS), hydrogen, and advanced biofuels, where [major oil and gas companies] have placed the biggest bets in the energy transition" (Eberhart 2022). According to the energy analytics firm Enverus, "The new law signals [to oil and gas companies that] Democrats are willing to work with them and abandon the notion fossil fuels could soon be rendered obsolete. . . . Both supply and demand will increase over the next decade" (Brown and Phillis 2022; Dobbs 2022).

That major fossil fuel and financial corporations have embraced centrist climate policies, like the IRA, is not evidence of any sort of emancipatory awakening on their part, but rather a strategic public shift born out of shrewd self-interest. JPMorgan Chase CEO Jamie Dimon recently complained about the extreme agenda of environmentalists: "Why can't we get it through our thick skulls," he griped, in a call to shareholders, "that if you want to solve climate [change], it is not against climate . . . for America to boost more oil and gas?" (Towey 2022). According to one climate watchdog group, "the world's 60 largest commercial and investment banks . . . poured a total of 3.8 trillion into fossil fuels from 2016–2020," with JPMorgan Chase leading the way (Banking on Climate Chaos 2021). As Andreas Malm and the Zetkin Collective (2021) have argued, oil and gas companies have long been adept at speaking with forked tongues, touting the greening of their operations while focusing the lion's share of their efforts on increasing fossil fuel production. Koch-funded Americans for Prosperity unsurprisingly argued that the IRA would "raise prices and make American energy more expensive" (Schwartz 2022a). The American Petroleum Institute and a host of state-level oil and gas associations bemoaned the new taxes and alleged constraints on domestic energy production (API 2022). Home Depot cofounder Bernie Marcus's complaints about the Biden administration are par for the course:

> We are sitting on tons of oil and gas, and he doesn't allow us to drill. It's basically stupid. I can't come with another word. It's stupid. And we will continue to suffer this inflation. And our people will continue to struggle with high prices as long as he maintains this facade of climate control versus allowing us to do the things we should be doing. (Sozzi 2022)

Racing to the aid of the fossil capitalists, Republicans and conservative politicos stood united in venomous opposition to the IRA. Senate minority leader Mitch McConnell made the case that "the only things their 'Inflation Reduction Plan' will reduce is American jobs, wages, after-tax incomes, energy affordability, and new life-saving medicines" (McConnell 2022). Of the IRA, Missouri Representative Jason Smith wrote that "working-class Americans cannot afford it and should not have to live with it" (Smith 2022). Writing in conservative periodical *Newsweek*, a commentator called it "a slap in the face" and "a real gut punch" to the working class (Polumbo 2022). Noting that the IRA would "shower higher-income individuals with subsidies," a (Koch-funded) Mercato Center Fellow mused that the Democrats were "taking from the working-class [while] giving to the laptop class" (De Rugy 2022).

Fossil Fuels for the Working Class

While conservatives have long heaped rhetorical attention on the working class, this renewed attention is also indicative of a recent shift in GOP strategy. On March 30, 2021, Indiana Representative Jim Banks sent a memo to GOP leader Kevin McCarthy outlining a seemingly far-fetched possibility: that Republicans could "permanently become the Party of the Working-class." Banks began by noting that more mechanics and custodians provided political contributions to the Trump campaign in 2020 than to the Biden campaign, while more college professors, marketing professionals, and bankers donated to Biden. Perhaps the most striking statistics he offered, however, detailed the outpouring of political support from Wall Street into Democratic coffers. At such a moment, he suggested, "the vast majority of the Republican conference doesn't want to return to a GOP-era that neglects working-class voters" (Banks and McCarthy 2021).

The prescriptions that Banks proffered—building the wall, curbing the power of big tech, refusing "aggressive covid lockdowns," and embracing "anti-wokeness"—would do very little to actually help the working class, but he nonetheless homed in on a genuine weakness of the mainstream Democratic Party: "Democrats' agenda is now shaped entirely by corporate interests and radical, elite cultural mores, but they still rely on many blue-collar voters." Of course, neither Banks nor his Republican Party colleagues demonstrate an inkling of concern for the ways in which environmental degradation is affecting the lives of these blue-collar voters, but they do present an alternative vision of the relationship between working-class people and environmental politics—one in which the Left is destroying the working class through its "war on fossil fuels."

The Banks memo is no mere outlier. Building on the momentum of Trump's appeal to extractive workers, the anti-environmentalism of the Right is coming to converge with a brand of conservative politics tethered to the purportedly "traditional" social values of a *multiracial working class* (Ruffini 2023). As former Nevada attorney general Adam Laxalt argued in his 2022 campaign for US Senate,

> What we've seen in the last few years are these Democrat policies, whether it's attacking energy independence or spending too much money causing inflation, is absolutely killing the working-class. . . . We're seeing conversions at all-time highs in this race [with] Hispanic working-class [voters], in particular. (Schneider 2022)

Unlike the typical conservative appeal to US workers over the past forty years—which forged a linkage between the interests of the white male working class and those of capital—this emergent version of conservative class politics is connected to a fanatical opposition to "wokeness," which is rhetorically conjoined to both liberal elitists *and capital.* For example, a 2020 article in the *Journal of American Greatness* mused that there are "twin paths to socialism" in the United States today: "equity and climate change alarmism." "To stop socialism," the author concludes, "Americans must stand up to the alarmists that claim bigotry and fossil fuel are existential threats" (Ring 2020). The American Legislative Exchange Council, prodded by industry front groups like the Texas Public Policy Foundation, has set its sights on "energy discrimination"—prohibiting that state money flow to financial firms who seek to prioritize a so-called environmental, social, governance (ESG) approach to investing (Aronoff 2022a). "ESG is a scam," wrote Elon Musk (after Tesla was removed from the S&P's ESG index because of complaints of racial discrimination), "[that] has been weaponized by phony social justice warriors."[1] As a 2017 *First Things* article put it, "The most likely outcome of Biden's 'whole-of-government' response to climate change will be the expansion of America's post-industrial underclass." The author's conclusion: "Actions on climate policy and 'equity' illuminate the priorities of the Biden administration—and by extension, the Democratic establishment. The new administration has been prompt in promoting elite interests" (Berry 2021).

At the present moment, in which a renewed "racial reckoning" has again brought the need for racial justice to the forefront of public discourse, the ideology of whiteness is being renaturalized on the Right through fanatical opposition to politically correct "wokeness," so-called critical race theory, and environmental justice (even as the Right increasingly appeals to workers of color with conservative social values). It doesn't take much digging beneath the surface to see that this opposition to elitism and "woke capital" is, for most conservatives, far from a full-throated opposition to capital. As of 2022, the Republican agenda remains much the same as it has been: tax cuts, deregulation, austerity, and expansion of the military-industrial complex. For the Right, capital's cultural politics are the target, not its material interests. Put differently, the interests of fossil capital are being surreptitiously repackaged as part of a war on woke capital. In spite of the aforementioned rhetorical shifts, the general substantive thrust of conservative opposition to climate politics has remained essentially unchanged, going back to the earliest calls for environmental legislation; as conservative wunderkind Charlie Kirk recently put it, "Climate change is the wrapper around Marxism" (Media Matters 2022).

This representation of environmental protection as a Trojan horse for socialism highlights a deep continuity that has persisted in every class ecological order that I've traced over the course of the book. And yet, the present moment—where the Left is ideologically divided, capital is hedging its political bets while continuing to invest billions in fossil fuels, and the Republican Party is nearly united in its opposition to environmental protection but divided on its class politics—is one that differs in some respects from the previous periods that I've analyzed. We can make sense of these lines of continuity, structural instabilities, and ruptural potentialities by focusing on the evolution of two core concepts that I've introduced: *the natural wages of whiteness* and *the treadmill of social reproduction*.

THE NATURAL WAGES OF WHITENESS

In each historical era that I've analyzed, whiteness—the ruling-class social control formation that emerges as many white workers come to privilege their *racial interests as white* over their *class interests as workers*—has been a site of socio-ecological struggle. The relationship between whiteness and the politics of nature is straightforward, if not simple. Capital needs nature, both in the form of natural resources that can be channeled into the process of production and of sinks for the by-products of production. In the process of obtaining access to rivers, lakes, aquifers, land, forests, and minerals, capital almost always encounters resistance from the people whose lives are being upended by the accompanying dispossession and pollution emanating from processes of extraction, transportation, and production. In order to maintain possession of this actually existing nature over the resistance that occurs, capital needs a working class that is divided. Much of this division, as I've detailed throughout the book, comes from strategies of "job blackmail" that capital spreads like the gospel. But the reach of these messages is broadened enormously by the division that Du Bois long ago laid out—between white workers and workers of color. A vital component of the wage that entices people *who happen to be white* to become *white social and political subjects*—that is, to view their racial identity as central to their sense of self and society—is environmental. There exists both a material dimension to the natural wage of whiteness and an ideological (or what Du Bois terms "psychological") dimension. The former has historically appeared through privileged access to private and public lands, the ability to distance oneself from the worst ravages of industrial pollution, and the ability to live in close proximity to environmental amenities, like parks and open spaces. The latter has functioned through white workers' internalization of dominant ideologies of nature (wilderness, ecology, the environment, sustainability, etc.), which are so powerful (and insidious) because they offer recourse to a foundation beyond politics through which social hierarchy can be legitimated. This general pattern has nonetheless taken different forms over time.

After the Civil War, with the pro-slavery Democratic Party in tatters, African Americans, abolitionists and so-called scalawags and carpetbaggers enforced Reconstruction with the power of the state. Access to nature via both land and usufructuary rights to the commons occupied an integral place within the broader project of Reconstruction and the cross-racial alliance that enabled its short-lived successes. Reconstruction was always imperfect—failing to consider the rights of Native Americans whose dispossession paved the way for the promise of the wild frontier, and predicated on a precarious alliance with Northern capital. The promise it did contain was ultimately undercut by the strategic maneuverings of capital, accompanied by a resurgent cross-class alliance between whites. The raging renaturalization of whiteness proceeded, in part, through returning land to plantation owners and the strict interpretation and enforcement of private property rights via enclosure and fencing laws. These policies were legitimated by appeals to a combination of two emergent scientific ideologies: conservation and biological racism.

Amid the advances of the mid-twentieth-century civil rights movement, whiteness was again challenged as social movements sprung to the fore and unions began to glimpse the real promise of a multiracial industrial democracy. However, the state-subsidized rise of the über-segregated suburbs created spatial distance between both workers and environmentalists and people of color and whites, the latter division reinforced not by biological racism but by an environmental differentiation between culturally defined groupings. Among middle-class whites, the suburbs became a site of intense environmental organizing but also birthed a conservative countermovement, founded on a libertarian ideology hostile to environmental regulations, that would gain momentum in the decades to come. Of course, the financial support for this nascent anti-environmental movement came from on high—the Republican Party, the National Association of Manufacturers, the Business Roundtable, the extractive sector, and a growing network of conservative think tanks whose existence depended on ample funding from fossil fuel interests.

While its dollars were delivered from corporate boardrooms and its broad base of support remained in the suburbs, this countermovement would get legs only through its association with rural environs and the white male extractive workers who came to function as the vanguard of anti-environmentalism. Over the course of the late 1970s through the 1990s, a conservative lexicon of class emerged that pitted a "new class" of environmental elitists and an "underclass" of undeserving poor against a "working-class" comprising miners, loggers, roughnecks, fishermen, ranchers, and farmers. Such a project took hold in the Sagebrush Rebellion and Wise Use movement, which drew inspiration from the frontier mythology of the nineteenth century; it was ultimately solidified in US conservative dogma in fights over climate change. From "drill, baby, drill" to "the Silent Majority stand with Trump," this anti-environmental conception of the working class has, in recent years, surged to the forefront of US conservatism. American conservatives are attempting to resurrect the working class as the purview of all those who prize the national values embodied by extractive laborers. In the midst of this MAGA-class politics, the lines between Far Right white supremacists and the mainstream of the Republican Party grows blurrier by the day. White supremacy has reentered the mainstream of US political life, and—as I detail below—found a ready-made man camp in the fields of fossil fuel production (Nobel 2020).

At the same time, the politics of whiteness have grown increasingly vexed. As the discussion above suggests, whiteness has never been ideologically secure, much less logically coherent. But, over the past several decades, three phenomena have converged to further complicate the parameters of whiteness in the United States. First, more so than in previous periods, whiteness is today refracted through the terrain of US party politics. Racial attitudes are being filtered through prisms of ideology and partisanship, with white conservatives and Republicans being more likely than others to exhibit high levels of racial resentment (Enders and Scott 2019) and ethnic antagonism (Bartels 2020). Racially coded tropes long deployed by the Right in its fight to dismantle the social safety net (e.g., "welfare queens") have also spilled over into areas like environmental politics. Benegal (2018), for example, finds that whites with high levels of racial resentment are more likely to be climate deniers, while McCright and Dunlap (2011, 2013) conclude that there exists a strong correlation between conservative white men and anti-environmental attitudes. Recent (anti)environmental struggles—like the rage-filled response to the Oregon cap-and-trade bill that I described in chapter 3—also lend empirical support to scholarship suggesting that whites with a strong sense of racial solidarity are more likely to turn toward authoritarianism (Jardina and Mickey 2022).

Second, however, wages—in both the narrow economic sense and the more expansive sense employed by Du Bois—have seen a universal decline across the working class. At a time where real wages have been stagnant for decades, costs of living continue to mount, and "deaths of despair" have surged among whites without college degrees, in particular (Case and Deaton 2021)—the wages of whiteness don't buy what they used to. This includes access to environmental amenities (like land and natural resources), and the ability to distance oneself from environmental harms (like air and water pollution, and the impacts of climate change). The combination of pervasive material insecurity and increasingly prominent ideological linkages between environmental action and social justice on the political Left has cleared a pathway that environmentalists might follow out of the morass of whiteness. While there remains enormous work to be done, there are signs of forward momentum; from anti-pipeline mobilizations to rallies for climate justice to the fight for a Green New Deal, a growing number of white Americans have, in recent years, stood in solidarity with workers of color and indigenous communities who are fighting against socio-ecological injustices (see, e.g., Grossman 2017; Gedicks 2018; Estes 2019). As political theorist Cristina Beltrán observes, "We are seeing a growing rupture between white citizens who support the politics of whiteness and those white citizens who are increasingly averse to racist and xenophobic appeals to white supremacy" (2020, 121). It is worth recognizing that white liberals and even progressives are notoriously fickle when their ideological commitments run up against their personal and familial privileges—what Corey Robin calls "disturbances in the private life of power" (e.g., proposals to integrate schools, forge more inclusive zoning regulations, or change school curricula)—but the aforementioned struggles point to real political shifts at the interface of ideology and party that render majoritarian coalitions centered around racial, gender, and class justice increasingly realizable. In short, we live in a time where whiteness has grown both increasingly frightening and increasingly fragile.

Third, this has all occurred at a moment in which profound institutional barriers exist to challenging both social inequality and environmental degradation. With little success in building power through unions, parties, or state institutions, progressives have found some semblance of solace in carving out micro-level loci of institutional clout in nonprofits and particular corners of the academy (the humanities, arts, and social sciences, in particular). Here, driven by a necessary rage against the frightening resurgence of white nationalism but working within institutions populated by professional class liberals and progressives that have little in the way of connections to working-class life, the fight against structural inequality has shifted to the terrain of culture (with emphasis placed on dismantling discrimination via the representational politics of language and aesthetics). Whereas, in the 1960s, radicals in Students for a Democratic Society urged factory workers who they toiled alongside to recognize their "white skin privilege" and revolt against both capital and white supremacy (see, e.g., Ignatin and Allen 1969), the anti-racism of today's professional class bemoans the interpersonal and intraorganizational vestiges of an all-consuming "white supremacy culture" and omnipresent "white fragility" via in-house trainings on diversity, equity, and inclusion (DEI) led by well-compensated corporate consultants. I am not suggesting that such initiatives are altogether unwarranted—in many cases they emerge out of a necessary need for organizational reflection and change—but insofar as they serve as a proxy for political engagement that in previous eras might have occurred within parties, movements, and unions, their promises of social transformation are chimerical. Moreover, there is growing evidence that, absent the organizational channels or grassroots outreach that might build bonds of solidarity with working-class communities, this liberal approach to DEI is filtered into popular consciousness as a form of elitist virtue-signaling that increasingly alienates workers across lines of race and gender.[2]

Herein lie the makings of a tragic paradox: at a moment when white supremacy is surging on the Right, and the wages of whiteness are imperiled by the growing specter of cross-racial solidarity on the Left, whiteness is being renaturalized by ostensibly liberal forces in ways that could undercut the mass movements and majoritarian coalitions that lie within our grasp—thus undermining the potentially transformative political projects that we so badly need to succeed (see, e.g., Mitchell 2022; Táíwò 2022). In the cracks and fissures of whiteness lie strategic opportunities for those concerned with social and environmental justice. The pressing question, however, is this: Given the minoritarian realities of US political institutions—the antidemocratic institutional rules and procedures that often prevent the Left from turning grassroots energy into concrete political power—do their exist pathways that an emergent working-class environmentalism could follow into socio-ecological transformation? In the section that follows, I make the case that, although the barriers are considerable, understanding the shifting relationship between the treadmill of production and what I've termed the treadmill of social reproduction might at least help lead us down the right trail.

THE TWO TREADMILLS: PRODUCTION AND SOCIAL REPRODUCTION

The relationship between environmental politics and class struggle is mediated by two treadmills—of production and of social reproduction. As I detailed in chapter 2, the treadmill of production emerged to describe how the metabolic relationship between nature and labor had been transformed in the mid-twentieth century by an energy- and chemical-intensive form of production that substituted greater swathes of nature (as resources used as inputs into the production process and sinks into which waste is expelled) for fewer workers. The ToP thus calls attention to the dialectical relationships between both *capital and nature* and *capital and labor*. The treadmill of social reproduction, by contrast, refers to the ways in which care work—cleaning, cooking, procuring food, caring for children and the elderly, and maintaining the home and its environs—also shifted in relation to the aforementioned changes in production. By moving our focus "beyond the hidden abode of production," as Nancy Fraser would have it, the treadmill of social reproduction complements the perspective offered by ToP theorists in highlighting how the labor necessary for the replenishment and continuation of collective life itself has been fragmented along lines of race, class, and gender.

Why is this complementary perspective important? In *The Great Transformation*, Polanyi argued—rightly, I believe—that as the commodification of land, labor, and money upends the socio-ecological relations necessary for habitation, countermovements will inevitably attempt to re-embed the market within society. As Polanyi knew well, however, society is not monolithic; the countermovements that emanate from the ruptures facing propertied elites are a far cry from those that rise out of settler colonial dispossession, racial injustice, sexist discrimination, or working-class impoverishment. These divergent modes of exploitation and oppression are filtered through class, racial, and gendered divisions of labor in both the workplace and the home, resulting in political mobilizations that differ in their demographic composition and substantive content.

As ToP theorists point out, the mid-twentieth century saw an increasingly salient division emerge between the lives that members of the polity lead as workers versus those they lead as "citizen-consumers." In the nineteenth century, the division between the domestic and public sphere was institutionalized; in the mid-twentieth century, it found its logical endpoint in the suburbs. At this moment, the double movement sprang into action, but it was divided: from the workplace came the chants echoing out of union halls and off of factory floors, calling for better wages and working conditions, and in the suburbs came the stomping feet of canvassers and petition-writers demanding the stabilization of the local natural and cultural environment. On one side, unions waged valiant fights in industrial cities against a treadmill of production that was going faster and faster. On the other side, housewives in suburban bedrooms ran in place on store-bought treadmills, struggling to stabilize the local environment against the threats encroaching from the world beyond. Industrial labor and suburban environmentalism were certainly not the only (or even the most influential) political mobilizations of the era—from urban centers, Native American reservations, and agricultural fields emerged the stirrings of rebellion. Rather, the politics of the suburbs are illuminating insofar as they provide insight into the dominant thrust of a middle-class ecological order. The suburban citizen was plagued by a split personality of sorts: their contradictory location within class relations (Wright 1980) enabled a lifestyle environmentalism to flourish alongside a more foundational commitment to property rights, economic growth, and consumer pleasure (Huber 2013).

By the late twentieth century, as social reproduction had become increasingly commodified—that is, integrated into a global treadmill of production—the pressures on the forms of labor occurring at the site of both production and social reproduction were increasing for the vast majority of residents of the United States. On one hand, the "compromise of embedded liberalism," in which labor won real concessions during a period of sustained growth, was falling apart amid the ruptures of neoliberal globalization. On the other hand, the "family wage"—in which a bread-winning patriarch could provide a middle-class lifestyle for his entire family—was no more. Rather than seeking to re-embed the market within society via a social democracy that was more inclusive and democratic than the Keynesian welfare states of the mid-twentieth century (i.e., one that provided universal benefits to all), both major political parties turned to a combination of financial capitalism and state coercion to deal with the aforementioned crises. Republican and Democratic commitments to neoliberalism differed in degree rather than kind: Republicans sought to dismantle the state regulatory apparatus in all but its most violent forms; Democrats sought to use markets in the service of welfare and environmental protection, while attempting to lure middle-class suburban voters by supporting the carceral state.

Amid the decimation of unions and the rightward tilt of the Democratic Party, the terrain of political struggle on the Left shifted decidedly away from production and toward social reproduction—to the home, neighborhood, and local community, with environmental justice emerging as perhaps the most emancipatory movement of the era. Through this form of "livelihood environmentalism" (Huber 2019, 2022), grassroots activists were able to win notable local victories in response to egregious injustices, but the general trend was one in which inequality began a dramatic acceleration that has continued ever since: real wages have been stagnant since the late 1970s while rates of unionization have continued their long decline; costs of living have mounted as housing, childcare, water, and agricultural land have become sources of financial speculation and institutional investment; the ubiquity of industrial pollutants has rendered massive swathes of soil dangerous to grow food in, water toxic to drink or bath in, and fish unsafe to eat; extreme weather has become common in many places, while others are quickly becoming uninhabitable. In short, the past four decades have witnessed a hollowing out of the social and ecological functions that have historically made *the oikos* home. Without robust unions that could build solidarity and combat capital in the workplace, without a party willing to fight for the rights of workers in the political sphere, and with local organizers fighting a one-sided battle that they aren't positioned to win, it is no surprise that social ties crumbled and many people turned inward, seeking solutions through self-help, self-care, and self-promotion, or instinctively clinging to the few forms of kinship that remained (Silva 2019).[3]

Much like the tragic derailment in East Palestine, Ohio, the train of improvement has sped off the tracks, leaving the surrounding habitation engulfed in toxic plumes. The late twentieth- and early twenty-first-century dreams of emancipation via eco-localism have faded into nightmares of ecological destruction and purgatories of toxic existence. Pessimism rules the zeitgeist, absent the optimism of the will that animated Gramsci's oft-invoked phrase. The hour is certainly late, but if you squint hard enough at the night-time sky, it's possible to see glimpses of two movements that once possessed world historical consequence—socialism and fascism—coming again into our orbit.

THE STRUGGLE TO COME

In the aftermath of the January 6 insurrection, fascism surged back into US political dialogue after an almost eighty-year hiatus. Among left-leaning academics and public intellectuals, debate raged over whether Trumpism represented a resurgent neo-fascism fanning the flames of blood and soil (Burley 2017; Katz 2021; Paxton 2021) or a more quotidian form of minority rule long baked into the structure of US politics (Riley 2018; Robin 2021). While revealing genuine interpretive differences among those who have staked out positions, the debate presents a false choice between the exceptional outburst of fascism and the normal terrain of US politics. As Polanyi long ago argued, widespread frustration and immiseration caused by an inability to check the commodification of life creates the conditions of possibility for fascism. To the real extent that minoritarian institutions and procedures (e.g., the Senate, federal courts with lifetime appointments, gerrymandering, etc.) reinforce this barrier, they contribute to the neo-fascist project. The unchecked, mundane violence of our existing order and the sumptuous, explosive violence of fascism exist along a spectrum; the inability to stop the former builds momentum for the latter. The fact that the jokers of January 6 were incapable of actually seizing power does not mark a grand break between existing fascist potentialities and historical realities: "In no case was an actual revolution against constituted authority launched; fascist tactics were invariably those of a sham rebellion arranged with the tacit approval of the authorities who pretended to have been overwhelmed by force" (Polanyi 1944, 246). Moreover, the relative weakness of existing Far Right organizations in comparison with their historical predecessors is not as much of a reason for relief as we might like to believe:

> A country approaching the fascist phase showed symptoms among which the existence of a fascist movement proper was not necessarily one. At least as important signs were the spread of irrationalistic philosophies, racialist aesthetics,

anticapitalist demagogy, heterodox currency views, criticism of the party system, widespread disparagement of the "regime." (Polanyi 1944, 246)

In the aftermath of the Capitol insurrection, the chances that US governmental authorities, backed by a powerful fraction of capital, might accede to the violent longings of Far Right antidemocratic forces in the not-too-distant future seem all too real.

Fossil Fascism

In their 2021 book *White Skin, Black Fuel*, Andreas Malm and the Zetkin Collective provide a compelling picture of how minoritarian institutions and grassroots momentum on the Far Right might combine to create the conditions of possibility for fascism to reemerge in a world on fire. Malm and the ZC define fascism as "a politics of palingenetic ultranationalism that comes to the fore in a conjuncture of deep crisis, and if leading sections of the dominant class throw their weight behind it and hand it power, there ensues an exceptional regime of systematic violence against those identified as enemies of the nation" (2021, 235). The politics of *palingenetic ultranationalism*—the sense that the nation is dying and needs to be reborn (Griffin 1993)—is readily alive today on the US Right, but what of the conjuncture of deep crisis and the support of the dominant class? Fascism rode into power on a tidal wave of explosive immiseration, economic dislocation, and national embarrassment that exceeds anything being experienced by the masses who are flocking to right-wing parties across much of the world today (certainly in the US and EU). World War I, as Clara Zetkin described it, "shattered the capitalist economy down to its foundations." The COVID-19 pandemic, in spite of its profoundly disruptive impacts, did not do this. And while climate change has already produced enormous suffering, displacement, and death, it does not yet threaten the accumulation of capital in a serious way. Will the "slow violence" of climate change and toxic exposure ultimately grow into a crescendo of world historical rupture?

Building on the sociopolitical and ecological patterns of the past and present, Malm and the Zetkin Collective provide two potential scenarios in which climate change could come to intersect with the contradictions of capitalism and existing forces of proto-fascism to produce fossil fascism as a mass movement. The first is a *mitigation crisis*, where successful efforts to respond to climate change—in the form of, say, a Green New Deal as envisioned by Democratic Socialists—pose an immediate and existential threat to the fossil fuel industry, petrochemical manufacturers, and industrial agriculture: "The assets buried in the seams will be irretrievable, mouth-watering business opportunities, fixed capital of mammoth size, an entire department of accumulation will be forever gone" (2021, 240).

At such a conjuncture, primary fossil capital, fighting for its survival, might be pressed to throw the full force of its weight behind the Republican Party and its increasingly radical networks of grassroots support, demanding a cessation of climate mitigation by extra-juridical means. We see signs of such a social formation already: oil and gas companies are actively involved in the business of violent repression, pushing "critical infrastructure bills" that criminalize pipeline resistance, and channeling funds to private security forces working to quash anti-pipeline protests (Brown 2018). There is, today, a sizable presence of Three-Percenters in the Bakken oil fields (Nobel 2020), and the biggest billionaire donor to Trump's 2020 campaign was none other than Kelcy Lee Warren, of Energy Transfer Partners, which owns a major stake in the Dakota Access Pipeline (Collins and Ocampo 2021). The potential for a mitigation crisis seems very real.

The second pathway for fossil fascism to rear its head is an *adaptation crisis*. "Imagine," Malm and the ZC ask of their reader,

> that repeated climate shocks chip away at the material foundations of societies at a level far deeper than in the first two decades of this century: heatwaves five or ten degrees hotter; wildfires roaring through regions for months on end; food provisioning systems at breaking point; storms pushing the sea dozens of kilometres inland—here there is little need to exercise the faculty of the imagination. It is portended in the science. (2021, 241)

The impacts of a warming climate will cause particularly acute disruption in the poorest areas of the world and will almost certainly provoke mass migrations. These migrations will be confronted with both the deeply embedded xenophobic and racist narratives that are surging on the Right today and the hollow centrist solutions that have been proffered by liberal politicians, like the Obama and Biden administrations—for example, paths to citizenship wedded to border securitization absent any systematic rethinking of US imperialism. In such a scenario, it is likely that that fossil capitalists will embrace superficial "solutions" to climate change, such as calls for population stabilization via immigration restrictionism—"border walls gone green," as I've called it elsewhere (Hultgren 2015). Political theorist Cara Daggett worries that, faced with both the loss of a political home and the social and ecological disruption of the *oikos*, "climate change could catalyze fascist desires to secure a lebensraum, a living space, a household that is barricaded from the specter of threatening others" (2018, 44). While most Far Right groups are militantly pro-fossil fuel, there is a generational divide in hypernationalist movements, and many Far Right parties and figures across the world are flirting with various forms of eco-nationalism (Malm and the Zetkin Collective 2021). In a world on fire, the specter of threatening Others—not only immigrants but socialists, LGBTQ people, and racial justice, indigenous rights, and environmental activists—could provide the scapegoat that fossil capitalists need to legitimate their counterrevolutionary longing for business as usual.

But the fossil capitalists won't be able to actualize these projects themselves; their oil-soaked dreams depend on cross-class support from workers. Amid the tumult of the inter-war period, Zetkin observed that fascism was rapidly becoming "an asylum for all the politically homeless, the socially uprooted, the destitute and disillusioned. . . . And what they no longer hoped for from the revolutionary proletarian class and from socialism, they now hoped would be achieved by the most able, strong, determined, and bold elements of every social class" (Zetkin 1923). In *The Origins of Totalitarianism*, Hannah Arendt similarly remarked on how fascists took advantage of a disorienting sense of alienation with status quo political leaders that had emerged post–World War I:

> The fall of protecting class walls transformed the slumbering majorities behind all parties into one great unorganized, structureless mass of furious individuals who had nothing in common except their vague apprehension that the hopes of party members were doomed, that, consequently the most respected, articulate and representative members of the community were fools, and that all powers

that be were not so much evil as they were equally stupid and fraudulent. (1973, 315)

Conservative political strategies that emerged in response to COVID-19 were geared to take advantage of this sense that the ground beneath our feet had become shockingly unstable—that those without a political home could find one in the American nation. As the usual relations of daily life melted into thin air amid a global pandemic, we witnessed borders solidify with an astounding alacrity. The white supremacist American renaissance crowed that "borders are suddenly back in fashion everywhere" (Taylor 2020). Pat Buchanan (2020) giddily suggested that this confirms a timeless truth about humanity: "When a crisis comes, be it a war in which the survival of the nation is at stake or an epidemic where the health and survival of our people is at stake, we take care of our own first." Rich Lowry (2020) of the *National Review* bemusedly bragged that the nativist Right was right all along: "We are all restrictionists now."

As the immediate urgency of the public health threat became filtered through social and popular media, and as social-distancing measures stretched on over months, right-wing posturing coalesced into outbursts of organized rage against COVID-19 lockdowns, opposition to popular mobilizations against racial injustice, conspiratorial warnings about radical instigators (i.e., ANTIFA) with revolutionary intentions, and—of course—amped-up support for border walls, pipelines, and oil and gas extraction. These diverse grievances congealed into a set of socio-ecological principles that revolved around protection of the home in a world of instability. To protect our home, the narrative went, oil needed to flow, Americans needed to get back to working and socializing together, and immigrants and social justice activists needed to stay out of sight. It seems clear that whatever future right-wing responses to the climate crisis might look like, the state of emergency beckoned by fossil capital and its right-wing allies will certainly not be framed as being in the interest of fossil capital; rather, it will be pitched as in the interest of US workers and American family values, with rhetorical emphasis placed on linkages between the home and the homeland, local vitality and national renaissance, the stabilization of "our" social and ecological environment, and the ruptural threats posed by the Right's favorite boogeyman: socialism.

Eco-Socialism

The environmental Left has long argued that, at least in this one respect, the Right is right: any effective response to climate change necessitates a dramatic restructuring of the capitalist political economic order (Klein 2015). Is there reason to believe that political momentum is building for such a restructuring? Are there conditions under which the ecological crises of which Malm and the Zetkin Collective speak—coupled with intensified inequality and continued party polarization—might instead work to prod Americans in the direction of eco-socialism? The evidence is decidedly mixed.

Although the COVID-19 pandemic did not shake the foundations of global capitalism, it did provoke a temporary surge of state social spending that harkened back to the New Deal of the 1930s, as well as several explosions of collective outrage that temporarily channeled a series of intertwined grievances stemming from the material conditions faced by many—racist and classist policing, economic precarity, and social anomie—into a protean political form. This took the mold of nationwide #Black Lives Matter demonstrations in the aftermath of the murder of George Floyd by Minneapolis police officers (in close proximity to the murders of Breonna Taylor in Louisville and Ahmaud Arbery in Georgia); small but significant surges of labor organizing in education, health care, and logistics; and continued opposition to oil and gas pipelines and mines. As Keeanga-Yamahtta Taylor put it during what was, in hindsight, the high-water mark of the #BLM protests, "We're seeing the convergence of a class rebellion with racism and racial terrorism at the center of it. And in many ways, we are in uncharted territory in the United States" (*DemocracyNow* 2020). While the import and impact of Black Lives Matter, in particular, should not be underestimated, its considerable grassroots energy failed to translate into a mass organizational form that could be sustained, or to transform the Democratic Party's priorities beyond relatively narrow pledges of support for reforms in the courts and policing (Johnson 2017, 2023). This barrier is not unique to BLM but also plagued climate justice, Occupy Wall Street, and anti-globalization movements before it. How might what have come to seem like almost routinized outbursts of anger and opposition on the Left be translated into mass movements and majoritarian coalitions?

On the one hand, over the past decade, positive perceptions of unions have increased markedly, the percentage of Americans with favorable views of socialism has crept up slightly, and membership in the Democratic Socialists of America has risen to nearly ninety thousand members, which includes state and local officials across the country and five members of Congress (Hertel-Fernandez 2020). The closely aligned Congressional Progressive Caucus—which supports the PRO Act, Medicare for All, and a Green New Deal—has grown to include nearly half of all congressional Democrats.[4] As a consequence of our two-party system, an ideology that captures one of the two major party's bases actually gains access to power and, with it, the potential capacity to influence policy and appeal to the masses through the institutional channels of the state—with added attention from the media to boot. This is not to romanticize the two-party system or to suggest that activism should center around the Democratic Party or the state; rather, it is to recognize that socialist policies (Medicare for All, a federal jobs guarantee, etc.) once relegated to the margins are today entering the mainstream of public debate in a way that they haven't since the 1970s; the "Overton window" has shifted, and with it the prospects for socialism.

On the other hand, however, the socialist base—the working class—is as splintered as ever; indeed, there is evidence that the US working class has been bolting the Democratic Party, particularly in swing states that are vital to presidential elections and control of Congress. In this sense, the Banks memo is not altogether wrong. As numerous commentators have noted, the United States is in the midst of a *class dealignment*, in which poorer and less educated voters are shifting to the Republican Party and wealthier and more educated voters flocking to the Democratic Party (Karp 2023). This is particularly salient among white voters, but from 2012 to 2022, nonwhite voters without college degrees (an admittedly imperfect proxy for class) "shifted away from Democrats by 18 margin points" (Teixeira 2022; see also Teixera 2021; Walter 2021). Thanks to much publicized (and much-needed) struggles in the logistics (e.g., Amazon warehouses), education (e.g., grad students and adjuncts), and service (e.g., Starbucks) industries, there is a *feeling* on the Left that unionization is rebounding from nearly a half century of decay, but there is thus far little evidence that this is more than wishful thinking (Guastella 2022). In 2022, there were twenty-two "major work stoppages," involving a thousand or more workers, which is slightly more than the average for the past twenty years (sixteen/year), but far below the thirty-five/year average from 1990 to 2000, and nowhere near strike waves of the 1950s to 1970s (Bureau of Labor Statistics 2023). The 2022 unionization rate in the United States was the lowest it has ever been, at a scant 10.1 percent. And, with the exception of the 2023 United Auto Workers strike (which I address in more detail below), where labor struggles have surged to the fore, it has been in occupations and sectors, like education and health care, that are rarely aligned with the Republican Party or the politics of Trumpism. "While college-educated workers see labor unions more favorably than they ever have," writes labor activist Dustin Guastella, "the exact opposite is true for those without any college education." This is not altogether novel; as Kim Moody observed several decades ago, "The left with its highly theorized, often moralistic politics, and worker activists with an un-theorized pragmatic outlook are often like trains passing in the night" (2000, 2). In short, both Democratic Socialism and the form of organization that has historically served as both a springboard and support for socialist movements—labor unions—remain marginal presences in contemporary US political life. The Left continues to lack a mass movement with explicitly socialist politics, a sustained grassroots presence in working-class communities, or any sort of real traction in party politics. What, then, is the way forward?

In perhaps the most comprehensive strategic analysis of the linkages between climate change and class struggle to date, Matthew Huber persuasively argues that the first step should be to reorient our focus back to production, focusing "our organizing energy against the particular class fraction of the capitalist class that controls the production of energy from fossil fuels and other industrial carbon-intensive industries" (2022, 4). Whether this advances by unionizing the electric utilities, as Huber advocates, or by building on existing momentum in the auto and logistics sectors, organizing at the point of production is the most vital task of today's Left. But, given the aforementioned paucity of unionized workers today, this must be accompanied by complementary strategies that appeal to the day-to-day realities of working-class people. The clearest pathway, in my opinion, is to build Democratic Socialism into a mass movement that will (1) pressure the Democratic Party to endorse policies (the PRO Act, Medicare for All, and environmental proposals like the Sanders GND) that promise real material benefits for the working-class; (2) work with unions and social movements, like environmentalists, #BLM, and indigenous rights activists, to build both grassroots support and associational ties not only in deep blue cities and suburban enclaves but also in working-class communities in rural areas and red states; and (3) connect with the lived realities of working-class people by emphasizing the breadth of existing material inequality and deprivation—simultaneously economic and environmental—in our policies and narratives. Such a project has the potential to safeguard the *oikos* not through familiar recourse to the nation and the "traditional" family, but through collective resistance to the global political economic order forged through the creation of structures of care and bonds of solidarity. Integral to this project is the momentous task of ensuring not only that we rapidly transition to a post-carbon economy, but that we do so with justice for workers in mind.

A JUST TRANSITION FOR ALL

The concept of a Just Transition first emerged to advocate for the rights of workers amid economic transitions—away from warfare and away from particularly damaging forms of pollution. When Oil, Chemical, and Atomic Workers (OCAW) leader Tony Mazzocchi proposed a Superfund for Workers in the early 1990s, he was making a case for state intervention to proactively soften the blow of environmental regulations on laborers in chemically and energy-intensive industries who were already grappling with the harsh realities of neoliberalism. As labor advocate Les Leopold put it, "The basis for Just Transition is the simple principle of equity." Rather than asking workers to shoulder a disproportionate burden for environmental protection, the costs "should be fairly distributed across society" (Labor Network for Sustainability 2016, 6).

The term just transition is often used to refer to the short-term interests of fossil fuel industry workers—for example, support for coal workers amid climate mitigation—but in its more progressive iterations, its reach expands far further, aiming "towards a more socially and ecologically just Green Transition as an alternative to the various forms of green capitalism and green growth competing for hegemony" (Stevis and Felli 2016, 35). In this vein, it is worth remembering that when Mazzocchi first coined the concept, alongside Leopold, their original vision for a JT was significantly more radical than the understandings of many contemporaries who use the term. In 1985, the OCAW and Labor Institute proposed a Worker's Bill of Rights that included full pay, health care, daycare, and tuition for laid-off workers. Mazzocchi even envisioned a "University of Reclamation" where laid-off oil chemical industry workers would learn how to clean contaminated sites, in addition to getting a broader liberal arts education (Leopold 2007, 417–418). Such a vision is advanced today by groups like the Labor Network for Sustainability (LNS), who advocate for a JT that includes state funding for a wide range of initiatives: rapid response teams to address plant closures or mass layoffs, bridge funding for localities, wage replacement, alternative employment, health insurance coverage, childcare, relocation support, and continued pension and retirement contributions (2021, 61–62). LNS makes the case that these local-scale interventions must be combined with broader national level reform—a federal-funded Workers Superfund (à la Mazzocchi), the provision of a universal social safety net, passage of the PRO Act, and support for indigenous rights that includes free prior-informed consent.

By slowing down the treadmills of both production and social reproduction, the institutionalization of such a just transition would allow many workers to continue living dignified lives in their home communities. It is nonetheless clear that the short- and medium-term future is one in which many people across the world will be displaced from their current homes. In such a conjuncture, it would be a tragedy if the American Left pursued a strategy of eco-socialism in one country. As we work to end the imperial encounters that give rise to mass dislocation, any just transition worthy of the name will seek to make the right to freedom of movement for workers the rule rather than the exception. A major source of division within the working class that has historically aided the success of anti-environmentalism has been the differential mobility of working-class people—more specifically, the antagonisms that develop between those thrown out of place amid the ruptures of capitalism and those who attempt to salvage place-based roots of community and social ties amid similar pressures.

In each historical era that I've analyzed, particular segments of the working class have, in different ways, been fixed in place by a combination of the coercive force of the state, vehement local opposition to prospective migrants, or a simple lack of capital needed to finance mobility. The systems of exploitation and oppression wielded against, say, Native Americans violently forced onto reservations and coal miners channeled into tent colonies differed dramatically; the movement of these populations was restricted and policed by different actors, employing different strategies, and resorting to different forms of consent and/or coercion. But lurking behind these varied histories and trajectories is a strategic advantage of capital that both draws on and reinforces the natural wages of whiteness that I've described throughout the book. Stuck, quite literally, amid the ashes and soot of industrial-scale resource extraction—or, today, the rising tides and prolonged droughts of a climate-changed world—the place-based segment of the working class faces a real trade-off between the environmental harms from which they suffer and the jobs that they direly need. In an era in which nature is commodified and capital is free to flow across political boundaries of all sorts, restrictions on the mobility of the working class drive a wedge between the aristocracy of labor (who possess the political rights and economic wealth to insulate themselves, to some degree, from the worst of capital's ravages); the floating surplus populations of the world (e.g., the migrants who flee home in desperate pursuit of employment, security, and opportunity); and those either left behind or tightly tethered to the places that they love, who deal daily with the threats to habitation that follow in capital's miserable wake. This wedge appears ideologically in the form of a racialized nationalism; a vision of an America made great through higher wages for some, combined with walls and cages for others.

When the migrants caravanning north from Central America approach the US border holding signs proclaiming "we are workers," the American Left would do well to recognize them as such. When prisoners lament the toxic contaminants that they're exposed to in their sub-minimum-wage jobs, we should do the same. And when coal miners express concern that they might lose their jobs should climate policy be enacted, we should take pains not to dismiss their anxieties as mere stamps of support for Trump's white supremacy. Conversely, when fossil capitalists moan and groan about the excesses of a

Green New Deal, we should double down on its largesse. When they seek to bypass domestic environmental regulations by shifting production abroad, we should implement a border tax on the transnational movement of carbon and other greenhouse gases. And, of course, we should make the case for putting fossil fuel executives on trial for decades of profits forged through deception and violence; in the short term, their transition from gated communities to guarded cells will be a just one. Such a project—*freedom of movement for labor, walls and cages for capital*—could help to build bridges between workers across national boundaries.

In this more expansive sense, a just transition could be seen as a transitional demand—a condition of possibility for the more utopian task of political economic transformation. As Mazzocchi, Leopold, and organizations like the Urban Environmental Conference and Environmentalists for Full Employment long ago recognized, a just transition represents the bare minimum for the realization of a broader labor-environment coalition, a multiracial working-class environmentalism on the back of which any eco-socialist program will be built. JT, of course, is "a 'living concept' whose origins and meanings lie deep in the everyday experiences of workers and frontline communities" (Morena, Krause, and Stevis 2020, 2). The demands that constitute a just transition will shift as we segue from one conjunctural crisis to the next. Amid the throes of COVID-19, for example, the Climate Justice Alliance—a coalition of grass-roots labor, indigenous rights, and environmental justice groups—called for a "Just Recovery," which includes the provision of universal health care, paid sick leave and unemployment, a just transition to a post-carbon economy, and the direct participation of local communities in the creation of "the best care models for their most vulnerable" (Climate Justice Alliance 2020).

As COVID-19 has begun to wane, these political projects haven't stopped. The forces of "Blockadia" (Klein 2015) continue to press forward, but—as I reviewed in chapter 4—victories have been difficult to come by: "The current army of direct action eco-activists possess only limited disruptive capacity" (Huber 2022, 231). The divisions that have emerged between blue-collar workers and environmental activists—the former disproportionately working class, non-college-educated, and living in rural areas and the latter disproportionately professional class, college-educated, and living in urban and suburban areas—has been one significant barrier to the sustained success of protests attempting to thwart extractive projects, as well as the broader task of building a majoritarian coalition capable of mounting an effective campaign for transformative policies, like a Green New Deal. Clearly, protests and encampments aiming to stop oil and gas pipelines, export terminals, and dangerous mines are a necessary part of any eco-socialist political repertoire, but a strategy that centers around blocking extractive infrastructures leaves us reacting to fossil capital's strategic moves rather than proactively building power among the working class.

In this vein, the fall 2023 United Auto Workers strike against the big three automakers (Ford, General Motors, and Stellantis) offers reason for hope. In the face of record profits, auto workers had, over the past decade and a half, experienced stagnant compensation, a tiered-pay structure that disadvantaged younger workers, and the prospects of jobs and investment transitioning to a (nonunionized) electric-vehicle sector. When executives from the auto manufacturers refused to meet the workers' demands for ameliorating these conditions, 97 percent of workers voted to strike. While Biden voiced support for the union (becoming the first sitting US president to visit the picket lines), Trump made repeated pitches to the workers, blaming environmentalists for their plight. "The UAW workers," read one of his more intelligible tweets, are "being sold down the river by the union in favor of [a] green agenda." Capital echoed these sentiments. Reminiscent of the 1970s, when the National Association of Manufacturers, Chamber of Commerce, and Heritage Foundation attributed inflation to environmental regulations, auto manufacturers blamed environmental policy, claiming that in order to finance the costly transition to EVs while remaining afloat, they need to keep wages low (Ewing 2023).[5] UAW President Shawn Fain roundly rebutted this all-too-familiar "jobs versus environment" refrain: "We've said for months, we refuse to allow the EV transition to become a race to the bottom. . . . Corporate America is not going to force us to choose between good jobs and green jobs. That's a false choice" (UAW 2023).

Environmentalists, as well a wide array of social justice organizations, rallied to support the workers (Larsen 2024). The Labor Network for Sustainability played a major role in this cross-movement coalition, drafting an open letter to CEOs of the auto companies that was endorsed by more than 130 groups. "We are putting you on notice," the letter stated, that "corporate greed and shareholder profits must never again be put before safe, good-paying union jobs, clean air and water, and a liveable future" (Labor Network for Sustainability 2023). Representatives from LNS, Greenpeace USA, and the Sunrise Movement joined the picket lines. After forty-six days, the union had arrived at an agreement with all three companies. The workers won 25 percent wage increases, annual cost of living adjustments, and major reforms to the tiered-wage structure. What will perhaps prove most important in the long term, though, is that the new contract laid the foundations for unionizing the electric vehicle sector (Lichtenstein 2023). In a press conference announcing the agreement, Fain applauded the UAW members for their hard-fought victory, then appealed to the workers of the world, calling for a global general strike on May 1, 2028 (UAW 2023).

The roadblocks to actualizing a just transition away from fossil fuels and toward eco-socialism remain formidable, but the conditions of possibility exist for building a multiracial, working-class coalition that is structured, *in part*, around environmental demands—access to clean air, potable water, forests, rivers, lakes and parks, ample and healthy food, and a stable climate. Harkening back to Du Bois, political geographer Nik Heynen refers to such a movement for socioecological transformation as "abolition-ecology" (2018, 245). While the Right looks backward and yearns for a reinvigorated right-to-work nature that provides short-term economic concessions to predominantly white male workers in the extractive sector (at the expense of the health and well-being of these workers and their communities), we see glimpses of abolition-ecology in the coalitions working to contest pipelines and mines, in the UAW's current organizing campaign, and in the eco-socialist Left's vision of a revitalized working class—forged across lines of race, gender, and nationality—looking forward toward a worker and community-led Green New Deal (Cha et al. 2022) that is internationalist in its collaborations and aspirations for a better world.

Notes

INTRODUCTION

1. In Marx's general formula of capital (M-C-M'), he defines surplus value as "the increment" that distinguishes the money initially invested from the profit ultimately realized (1867 [1976], 251). This profit, which sets capital into motion, is made possible by the exploitation of labor-power; the laborer always produces more value than he or she receives in return. "The production of surplus-value," according to Marx, "forms the specific content and purpose of capitalist production" (411).

2. As Boggs noted in *The American Revolution*, "The Negroes, whom the radicals do not ordinarily think of as workers, form a large proportion of this working-class force which is usually considered as the revolutionary force, while the native-born whites who have been able to move up with every change in production are less and less inside the working-class force" (1963, chap. 1). We might add to Boggs's discussion of the actually existing working-class force, sizable segments of the Native American population, immigrant workers, and workers of color more generally.

3. The concept of a class ecology, as Stefania Barca explains, was first "theorized in the course of community meetings featuring left-wing scientists (Barry Commoner among them) that the [Italian] communists organized in Seveso with the intent of mobilizing local people against corporate and government cover up" of industrial pollution (2014, 13). Matt Huber has further developed a Marxist notion of "working-class ecology" in *Climate Change as Class War* (2022), which I engage with in chapter 4 and the Conclusion.

4. In formulating this concept, I draw on two seminal works of political thought: Carolyn Merchant's essay, "The Theoretical Structure of Ecological Revolutions" (1987) and Desmond King and Rogers Smith's "Racial Orders in American Political Development" (2005). Merchant argues that ecological revolutions—"major transformations in human relations with non-human nature"—are the products of tensions between "a society's mode of production and its ecology, and between its modes of production and reproduction" (265–256). At particular moments—for instance, during the colonization of the Americas in the fifteenth century and the rise of industrial capitalism in the early nineteenth—these tensions explode

into contradictions that destabilize the existing social and ecological orders, resulting in "new constructions of nature, both materially and in human consciousness" (273). Per Merchant, we've been living in the capitalist ecological order for roughly two hundred years now. And yet, while it is perhaps a period of relative structural stability glimpsed over the *longue durée*, it has also been one of enormous, world-altering upheaval. In this vein, Desmond King and Rogers Smith's analysis of distinct racial orders within US political development is instructive; each racial order is constituted by a struggle between "white supremacist" and "egalitarian-transformative" actors and institutions. Racial orders, they write, "are ones in which political actors have adopted (and often adapted) racial concepts, commitments, and aims in order to help bind together their coalitions and structure governing institutions that express and serve the interests of their architects" (75). Examples include battles between pro-slavery forces and abolitionists in the mid-nineteenth century, and segregationists and civil rights activists in the mid-twentieth. While I'm not fully in agreement with the institutional thrust of King and Smith's schema, their ideal-typical periodization of racial orders within US history provides insight into the evolving nature of political struggles and shifting political opportunity structures—that is, sites of vulnerability and stability within a particular institutional order. This orients our vision toward strategic points of leverage at which agents fighting for social transformation might target their activism.

5.   We might also note a third approach: site-specific case studies that focus on deep description—via interviews or ethnography—of particular anti-environmental strongholds, like Appalachian coal country, Pacific Northwest timber country, and the western rangelands. Insights from these studies are woven throughout my analysis.

6.   In terms of primary research, I spent time in four archival collections: the National Association of Manufacturers records at the Hagley Museum and Library, the Environmentalists for Full Employment papers at the University of Pittsburgh Library, the FitzGerald Bemiss papers at the Virginia Museum of History and Culture, and the James G. Watt papers at the University of Wyoming American Heritage Center. In addition to the archives of newspapers and periodicals, I also consulted several digital archives, such as the abolitionist periodicals housed in the Library of Congress "Chronicling America" collection, the *Working Man's Advocate* issues in the Internet Archive and the University of Illinois' digital collection, and UC-San Francisco's "Industry Documents Library." The ethnographic dimension of my research is admittedly limited, but I conducted interviews with twelve environmental activists, attended numerous local public hearings and town hall meetings related to PFAS contamination, watched livestreams of other public hearings on water pollution and a proposed mine, and attended one regional meeting on PFAS convened by the EPA.

CHAPTER 1

1.   Saidiya Hartman, for instance, observes that "in *Black Reconstruction*, women's sexual and reproductive labor is critical in accounting for the violence and degradation of slavery, yet this labor falls outside of the heroic account of the black worker and the general strike" (2016, 166). Kevin Bruyneel argues that, in his discussion of land as a potential source of liberation for the multiracial working class, Du Bois unintentionally relies on and reinforces a settler

colonial mindset that disables a consideration of solidarity with indigenous workers or a conception of land that engages with indigenous practices and traditions (2021). Nick Estes adds that, while Du Bois was an ally to Native Americans, he "could not make sense of the vexed history of settler colonialism, nor of the continued existence of Indigenous nations in the south and elsewhere" (2019, 211–213). My point here is less to critique Du Bois's scholarship—which emerged out of the pressing needs of a particular historical moment—than to consider how his framework could be adapted for the purpose of building more universal bonds of working-class solidarity today.

2. Titles of poems in *The Liberator* include "Evening Hymn to Nature," "Sonnet to Nature," "The Gospel of Nature," "The Gladness of Nature," and so on. One 1850 poem in *The North Star*, titled, "The World Is Full of Beauty," made a similar juxtaposition between nature and existing social structures:

> The sunny hills and valleys blush ripe with fruit and grain;
> but the lordlings of the palace still robs his fellow men.
> Oh God! What hosts are trampled amidst this press of gold;
> What noble hearts are sapped of life, what spirits lose their hold!
> And yet upon this God blessed earth, there's room for everyone;
> Ungarnered food still ripens, to waste, rot in the sun.
> For the world is full of beauty, as other worlds above.
> And if we did our duty, *it might be full of love*. (Anonymous 1850)

3. These policies of enclosure were accompanied by a bevy of discriminatory laws that included poll taxes, literacy tests, segregated schools, and convict labor leasing, as well as "enticement" and "false pretense" policies that made it difficult and dangerous for farmworkers to change jobs (Cohen 1991, 225–230; Foner 2007, 61–64).

4. Trina Williams estimates that nearly one-quarter of the contemporary US white population has ancestors who received land via the Homestead Act (Williams 2000, p. 8). Whether or not this ownership translated into asset-building is unclear given the aforementioned messiness of homestead claims and the precarity of continued property ownership for many claimants, but the sheer contrast between white and Black access to the benefits of the Homestead Act is striking. While the Southern Homestead Act of 1866 was explicitly geared toward providing land to the formerly enslaved, "very few homesteads were granted to black claimants" (Williams 2000, 10). See also Cox, "The Promise of Land for Freedmen" (1958) and Merritt, *Masterless Men* (2017).

CHAPTER 2

1. This is a position that Bemiss reiterated in letters and speeches. In a 1964 address to the Virginia state division of the Izaak Walton League, he mused, "Our early ancestors conquered the outdoors. Our more recent ancestors drew miracles of productivity from it. . . . This generation's most dramatic challenge is its ability to discipline itself in the use of this heritage through active concern for the qualities of our environment" (Bemiss 1964).

2. Bemiss served on the Perrow Commission, which recommended the option of "local choice" as a middle course between the abolition of public schools and their racial integration (see Grundman 1972, 40–60, 245–54; Bartley 1999 [1969], 109–110, 325). The commission's report bemoans the federal imposition of such a plan and makes its underlying goal crystal clear: "Under these recommendations, no child will be forced to attend a racially mixed school" (1959, 6).

3. "But far more important," he wrote to Bush, "is the solemn matter of these most fundamental principles which are covered in Jack Kilpatrick's editorial series. . . . Jack has attracted a great deal of attention in the great public service of developing this series" (Bemiss 1955; as cited in Grundman 1972, 89–90).

4. In a speech titled "The Future of Public Education in Virginia," Bemiss wrote, "The choice between integrated schools and no schools is a very, very tough one. I am sure I do not have to tell this group how wrong and unjust and unworkable integration is" (Bemiss nd).

5. As Kim Phillips-Fein details, "Between 1967 and 1976 the average number of workers on strike per year rose by 30 percent. In 1970 alone there were 34 work stoppages that involved more than 10,000 workers each" (2009, 153).

6. Silvia Federici (2019) points out that the concept also has non-Marxist origins (in the physiocrats, in particular). We therefore need to "dispose of the assumption that to speak of social reproduction is by itself to take a radical stance." My aim here is to analyze how the shifting relations between the terrains of production and social reproduction, propelled by the imperative of capital accumulation, impact environmental politics.

7. As Claudia Jones put in her seminal work, "An End to the Neglect of the Problems of the Negro Woman," "The super-exploitation of the Negro woman worker is thus revealed not only in that she receives, as woman, less than equal pay for equal work with men, but in that the majority of Negro women get less than half the pay of white women. Little wonder, then, that in Negro communities the conditions of ghetto-living—low salaries, high rents, high prices, etc.—virtually become an iron curtain hemming in the lives of Negro children and undermining their health and spirit!" (1949, 5).

8. In 1950, only 34 percent of women participated in the labor force, as opposed to 86 percent of men. By 1970, this had grown to 43 percent, while the male rate of participation had declined to 80 percent (Toosi 2002).

9. Ramey (2008, 55) estimates that women's household labor—cooking, cleaning, shopping, and caring for family members—increased slightly between 1940 and 1965 before declining markedly in the subsequent decade. "Non-employed women" spent 50.8 hours a week in home production in 1940 and 51.6 in 1965; "employed married women" spent 26 hours a week in home production in 1940 and 28.9 in 1965. My argument is that the treadmill of social reproduction also expanded beyond the household to duties related to community stability. As feminist historian Susan Ware notes, just two years before publication of *The Feminine Mystique* (1963), Betty Friedan had found that the foremost complaint of suburban

wives and mothers was that "their lives were too fragmented because of their varied activities." Ware reflects on this as follows:

> They participated in a dizzying range of volunteer activities, including starting cooperative nursery schools, serving on school boards, working against desegregation, founding mental health clinics, and organizing choral groups or theaters-in-the-round. Almost one third were active in local party politics. (1990, 291)

Taken as a whole, for many suburban women, the total hours spent on unwaged social reproductive labor increased markedly.

10. For example, "a mere 1,500 of the 186,000 single-family houses constructed in the metropolitan Detroit area in the 1940s were open to blacks. As late as 1951, only 1.15 percent of the new homes constructed . . . were available to blacks" (Sugrue 1996, 43).

11. While both working-class whites and racial minorities eventually found their way into suburbia, "as a rule, the place created by such groups was a place apart" (Kruse 2005, 244). On working-class suburbs, see Berger 1960; on Black suburbs, see Blumberg and Lalli 1966, and Farley 1970.

12. The first enlargement of whiteness refers to the integration of poor and working-class Anglo-Saxons amid the extension of suffrage to non-property-owning white men; the second enlargement occurred as Germans, Irish, and Scandinavians—who had immigrated to the United States in large numbers over the first half of the nineteenth century—became "white" in the decades following the Civil War (Painter 2010).

13. Massey and Denton find that "levels of residential segregation between blacks and whites began a steady rise at the turn of the century that would last for sixty years. . . . by World War II the foundations of the modern ghetto had been laid in virtually every northern city" (1993, 30–31).

14. This is not to say that nature went away—it continued to have a vital presence within environmentalism: "Imagining pollution less exclusively in terms of pathology, and more as a challenge to what was natural, helped empower lay activists to take up the fight against contamination" (Sellers 2012, 131).

15. In her 1966 book, *No Place to Play*, Margot Tupper describes the idyllic beauty of suburban life in the early 1950s outside of Washington, DC: "acres of untouched woodlands which were a refuge for children," a little creek, wildflowers, wildlife, and white dogwoods (1966, 18). But, eventually, the bulldozers came:

> One day my little girl . . . ran into the house shouting, "Mother, there's a bulldozer up the street. The men say they're going to cut down the trees." . . . Indeed the bulldozers did come! These huge earth-eating machines raped the woods, filled up the creek, buried the wildflowers and frightened away the rabbits and birds. . . . In less than a month the first of two hundred look-alike closely-set small houses rose to take the place of our beautiful forest." (19)

16. Oftentimes, the environmental concerns were quickly conjoined to anxieties surrounding the spread of "urban problems." In some cases, forms of no-growth zoning emerged in response to legitimate environmental problems but, in others, they merely provided a progressive language through which to preserve the racial and class composition of suburban communities (Ward 2019, 95). However legitimate the source of the anxieties, the relative weakness of the aforementioned environmental coalitions often failed to mitigate the regressive social impacts of no-growth policies.

17. In the early 1950s, every year saw more than four hundred strikes that idled more than a thousand workers, but by the end of the decade this number had declined to roughly two hundred a year (Bureau of Labor Statistics).

18. While unions continued to grow their membership in parts of the country and to wage influential strikes, the inroads they'd been making in the South and Sunbelt were arrested almost overnight: "The unions reborn in the New Deal would now be consigned to a roughly static geographic and demographic terrain, an archipelago that skipped from one blue-collar community to another in the Northeast, the Midwest, and on the Pacific Coast (Lichtenstein 2002, 118–120).

19. The continued racial division of labor was particularly clear in auto factories: "In 1960, Black workers made up 65 percent of the production workers at the Rouge plant but a mere 3.5 percent of the skilled workers. Even worse, Blacks made up 45 percent of the production workers at Dodge Main in Hamtramck but none of the 1,500 skilled workers" (Rector 2022, 95).

20. The disproportionate burdens faced by Native Americans as the nuclear industry developed warrants particular scrutiny—from extraction of uranium to testing of nuclear weapons to the storage of nuclear waste. With a half-life of 4.46 billion years, the acute toxicity of uranium-238 will persist in Native lands and waters for a longer time span than the human mind can fully comprehend (Voyles 2015, 11; see also LaDuke 1999). It is worth pointing out that uranium was mined not only by Native American workers; in 1950s and 1960s, 75 percent of miners were white, due at least in part to employment discrimination against Native workers (Voyles 2015, 13, 83–86; see also Malin 2015).

21. Treatments of race as biological nonetheless persisted. For example, a 1962 study of air pollution in the Detroit River basin sought to compare "the racial response to pollution" among different demographics, rather than considering the differential rates of exposure to pollutants experienced by these demographics (Rector 2022, 94–95; citing International Joint Commission, Report of the International Joint Commission on the Pollution of the Atmosphere of the Detroit River Area) (Washington, DC: International Joint Commission, 1962, 137–153).

22. The line of thinking present in the Moynihan report evolved from multiple scholarly and popular traditions that rejected eugenics and (biological racism more broadly) but reinscribed ethnic and racial boundaries on the terrain of culture. This included historian Oscar Handlin's "ethnic pluralism" framework (which Moynihan and Glazer cite approvingly in their influential *Beyond the Melting Pot*) (Reed 2020, 98–113), and the Chicago School of sociology's

"human ecology" approach (the Moynihan Report draws on the work of African American sociologist and Chicago school grad E. Franklin Frazier) (Reed 2008, 18–26; Browning 2022, 165–168). There is also a clear resonance with Oscar Lewis's "culture of poverty" theory (Taylor 2016, 36).

23. Robert Gordon points out that while the Sierra Club, Audubon Society, and National Wildlife Federation failed to support the Farmworkers, "several smaller organizations, including the Center for Science in the Public Interest, Environmental Action, Friends of the Earth, and the Environmental Policy Center, agreed to pressure the [EPA] to restrict hazardous pesticides and endorsed the UFW's boycott of California lettuce and grapes" (1998, 53).

24. Other environmental groups, like the Environmental Defense Fund and National Wildlife Federation, did ally with the Hopi in filing lawsuits to stop the extraction. See William Blair, "Indians Ask for Halt of Strip Mining," *New York Times*, May 15, 1971; see also Jack Waugh, "Smog over the Great Plains," *The Nation*, June 14, 1971, 753.

25. In 1969, a bill in the House of Representatives sought to establish a pollution disaster fund. Such a fund would have necessitated criteria for establishing what constitutes a pollution disaster area. As one pollution expert wryly observed,

    A pollution disaster area is defined in the bill as one in which the air or water is in immediate danger of becoming unsuitable or harmful for the uses traditionally made of it. By that definition, many of the Nation's large urban areas already are air pollution disaster areas, since the air in many of them already can be considered unsuitable for breathing and can be shown to be harmful to human health and welfare. (Esposito 1970, 7).

26. While the stereotype was one of urban Democrats moving to the suburbs and becoming Republicans, Bennett Berger points out that the partisan politics of suburbia varied significantly; for example, as steelworkers flooded into Levittown, Pennsylvania, the formerly Republican stronghold of Bucks County was transformed into a competitive district (29). Herbert Gans similarly describes healthy partisan competition in his ethnography of Levittown, New Jersey. The suburban demographic was a political prize for which both major parties vied.

27. One of the seven major objectives of the A. Phillip Randolph Institute's 1967 Freedom Budget was "to purify our air and water." Notably these environmental measures were tethered to full employment, as they also were at the UAW's 1976 Working for Environmental and Economic Justice and Jobs Conference, which I discuss in more detail in chapter 3.

28. The National Association of Manufacturers, for instance, opposed the Wilderness Act; proposed a new Department of Natural Resources within which the EPA would be (subordinately) located; suggested a range of tax credits to industry for pollution-reduction investments; opposed the ability of citizens to file lawsuits against industry for being harmed by pollution; sought to weaken the Clean Air and Clean Water Acts; and opposed regulation of sulfur dioxide emissions—to list but a few of their environmental policy positions.

29. In prescient ways, however, the article calls on the administration to push "for water pollution legislation that will significantly affect urban poverty."

30. *Danville Register*, "Agnew," September 22, 1972, pp. 1–2.

31. Of our environmental crisis, Bemiss (1970) wrote, "The blame for the present undesirable conditions cannot be fairly or productively put on enterprise or the free enterprise system. We are all to blame."

32. Windsor Farms, Inc., nd, "Windsor Farms, History," https://windsorfarms.org/history/.

CHAPTER 3

1. LASERs financial records were never made public, but—according to a local reporter—the group was funded by a number of large mining interests, including Mountain Fuel Suppliers, Rio Algom, and Kennecott Minerals (Graf 1990, 242–243).

2. The Mountain States Legal Foundation, for example, was founded in 1976 and first led by eventual Reagan Secretary of Interior James Watt. "I believe the big push for the MSLF," he wrote, in preparation for his interview with the organization's founder, Joseph Coors, "should be in guarding the natural resource base of America. . . . We need to establish a special expertise in the law concerning the energy issues—oil, gas, coal, uranium mining" (Watt 1977). Such a strategy was at least somewhat successful: "As of early 1979 there were 1809 lawsuits against federal laws, compared with 582 at the end of 1973; about 750 were against the EPA, 500 of them industry challenges" (Layzer 2012, 79).

3. Wacquant cautions against referring to this as mass incarceration for several reasons: less than 1 percent of Americans are incarcerated (not the masses), and if the masses were incarcerated, such a reality would have produced enormous political backlash (rather than remaining, until recently, the bipartisan default).

4. I realize that these are not one and the same—not all members of the New Democrat caucus or the Democratic Leadership Council adopted neoliberal ideals and not all neoliberals were members of the DLC; however, the Venn diagrams between so-called neoliberalism, New Democrats, and the Democratic Leadership Coalition contain enormous overlap.

5. Aronoff writes that "between 1980 and 2004, the fifteen countries initially included in the EU reduced SO2 emissions by 78 percent, compared to the US's 39 percent reduction over the same period" (Aronoff 2021, 78; citing Danish Environmental Research Institute 2004).

6. This early history of climate denial has been documented. Among the most thorough accounts are Oreskes and Conway (2010), Powell (2011), Turner and Isenberg (2018), and Rich (2019).

7. Powell notes that "one of the coalition's first efforts was to hire the same p.r. firm that pesticide companies had used to attack author Rachel Carson and her influential book Silent Spring" (2011, 94).

8. Letter to Reid and Boxer, June 6, 2008.

CHAPTER 4

1. For example, seven counties, two cities, two towns, and dozens of tribal governments and intertribal organizations in Upper Michigan and Northeast Wisconsin have passed resolutions opposing the Back Forty mine. Out of the eleven counties, cities, and towns, all but one were carried by Trump in the 2016 presidential election. At a 2018 DEQ hearing, three hundred concerned citizens packed a local high school gym—eighty-four residents testified against the mine, while four people spoke in support (River Alliance of Wisconsin 2018; Engel 2018).

2. A *Guardian* headline called climate change "the missing issue of the 2016 campaign" (Pilkington and Chalabi 2016). The *New York Times* noted the absence of climate change from the presidential debates, with *Media Matters* reporting that only 1 percent of debate questions were focused on climate change (Kalhoefer 2016). In the aftermath of the election, *E & E news* concluded, "Trump's narrow Electoral College victory, as well as exit polling, suggest *climate change and the environment played virtually no role in the presidential contest*" (Jacobs 2016; emphasis added).

3. The roots of this most recent iteration of the WWC can be traced back to the debates over the underclass that I examined in chapter 3. See, for example, Charles Murray's 1986 *National Review* article "White Welfare, White Families, 'White Trash.'"

4. Noting that whites of all education levels and age groups supported Trump, others turned to "whiteness" as the explanatory variable par excellence. Whiteness is not conceived here along the lines of Du Bois, Allen, or Ignatiev (though they are often cited in passing)—that is, as a ruling-class social control formation—but as one-time social construction that has morphed into a cultural, or even biological, essence (Taub 2016; Coates 2017; DiAngelo 2018). The *New York Times*' Amanda Taub (2016), for instance, defines whiteness as "membership in the ethno-national majority," asserting that whiteness achieves its power through the "privilege of not being defined as Other." As I outlined in chapter 1, I disagree with the more sweeping condemnations of whiteness studies in toto, but this now-dominant account of whiteness is a reification that obscures our understanding of the politics of actually existing white people, the terrain of contemporary inequality, and the forces producing today's reactionary politics. For those who find the new whiteness studies reductive, its purveyors have their own ready-made retorts—the white critics are beset by "white fragility" or blinded by "white rage," and the critics of color have unwittingly internalized dominant norms of whiteness. Whiteness here is both unfalsifiable and omnipresent—a ghost-like presence that pervades every social relation (and, indeed, every bit of reality); an ontological phantom that cannot possibly be excised from the *body politic* but can perhaps be managed via intrapersonal introspection and interpersonal genuflection. Reading the new whiteness literature, one is reminded of the critique that Oliver Cromwell Cox levelled against Gunnar Myrdal's influential analysis of US racism, *An American Dilemma* (1944). Liberal approaches to understanding racism, Cox argued,

> explain race relations away from the social and economic order. The theories do this in spite of the verbal desire of the author[s] to integrate [their] problem in the on-going social system. In the end the social system is exculpated, and the burden of the dilemma is poetically left in the "hearts of the American people," the esoteric

reaches of which, obviously, may be plumbed only by the guardians of morals in our society. (1945, 132)

5. The media also reported widely on the anti-environmental dimension of Trump country (see, e.g., Fisher 2016; *Washington Post* 2016; *Vice News* 2020).

6. The term "ecology" didn't come about until the mid-nineteenth century but builds on the work of Aristotle's student Theophrastus, who coined the term *oikeios topos* to refer to a "favorable place" where a plant grows best, its ecological niche. "It is probable," Donald Hughes writes, "that the classically educated nineteenth-century German scientists who coined [ecology] did so under the influence of the relevant passages in Aristotle and Theophrastus" (1985, 297).

7. In an interview with a journalist, Anahkwet expanded on this sentiment: "We, the Menominee, were given the responsibility to look after that river and land by the Creator thousands of years ago, and that supersedes any treaty or law. This Menominee River is a part of me; its essence is within my soul" (Rysavy 2017).

8. As Kelley notes, "During the late nineteenth century, for example, the militant strikes that erupted along the nation's railroads, in the mines of the Rocky Mountains and the steel mills in the East and Midwest, and among black washerwomen in the urban South were essentially community struggles. These strikes were never a simple matter of labor unions versus employers, and their success often depended on the fact that the local police, the families, and even some of the merchants sided with the unions" (Kelley 1999, 43).

9. She continued, "So for us, it's a matter of saying, 'no, part of the reason why these companies can pay such good wages, is that they're not actually incurring the true costs of what they're doing. They're not paying for the fact that they take people's water, they get subsidies from the state that allows them to inflate their wages. They should be paying you a lot more to take your coal, then what they actually end up doing, . . . because at the end of the day, if the bosses are being paid millions, you're not really benefiting that much. So for us, it's trying to explain that in terms of the myth that there are no jobs. There are jobs, it's just a matter of which jobs are considered more valuable" (Funk 2021).

10. The Service Employees International Union, National Nurses Union, Association of Flight Attendants-CWA, and the American Federation of Teachers have all passed resolutions endorsing Green New Deal proposals.

11. For example, conservative reactions to the Green New Deal have focused on threats to "traditional" lifestyles and habits—like meat consumption and automobile and home energy use (see, for example, Spect 2019).

12. In our interview, Anahkwet largely agreed with Gedicks's assessment but added a longer-term perspective: "I would say that it's got a good chance of being defeated, but . . . I know that as long as there's that mineral in the ground, there's always gonna be a threat. I kind of see [the struggle] as a lifetime thing, no matter what."

CONCLUSION

1. As Kate Aronoff (2022b) writes, "ESG is indeed a scam, but for essentially the opposite reasons of the ones Musk gave. There are still no agreed-upon standards as to what those three letters actually mean, despite the fact that $340 billion has flowed to ESG funds over the past two years."

2. For example, a 2021 study by Jacobin and the Center for Working-Class Politics found that

   > potentially Democratic working-class voters did not shy away from progressive candidates or candidates who strongly opposed racism. But candidates who framed that opposition in highly specialized, identity-focused language fared significantly worse than candidates who embraced either populist or mainstream language. (Jacobin Editors 2021, 5)

3. The academic Left has also turned inward—to theory detached from its material foundations (or, perhaps more precisely, to theory grounded in the narrowly circumscribed institutional realities of the professional managerial class). At a moment when workers across the world are seeking solace in those few social connections that haven't come undone amid the devastation of deindustrialization, deregulation, and austerity, segments of the academic Left have decided that the crises of the present herald potentially emancipatory futures: toxicity is queering our bodies, capitalism is destroying heteropatriarchal families, climate change is creating unruly, feral futures. Anthropologist David Bond notes, for instance, that many scholars are "converging on contamination as a physical rupture with the epistemic habits that underwrite modernity, as a kind of revolutionary release from the categorical reason that got us in this mess" (2021, 11). We might think of this as Shumpeterian identity politics: capital's depredation of life is accompanied by a creative destruction of modernist binaries. These particular iterations of new materialism and poststructural-tinged socialist feminism are fixated on linguistic acrobatics and a superficial aesthetics of transgression that ring horribly hollow—and are actively alienating—to the tens of millions who are left living with the weight of unsafe levels of exposure to carcinogens, who are grappling with the daily drudges of climactic instability, or who continue to find refuge in the love and care of their family (biological or otherwise) precisely because it's all that they have left. It is yet more evidence of the ascendance of an *antisocial socialism* that seems destined to snatch defeat from the jaws of a possible socialist resurgence (Guastella 2023).

4. For a list of current members, see the Congressional Progressive caucus Wikipedia entry, https://en.wikipedia.org/wiki/Congressional_Progressive_Caucus.

5. This, of course, is nonsense. The Economic Policy Institute has found that, since the 2008 financial crisis, the real wages of autoworkers have fallen by 19 percent, while big three auto manufacturers made more than 250 billion dollars in profits over the past decade (Hersh 2023).

References

PRIMARY SOURCES

ABC News. 2017. "President Trump Signs Energy Independence Executive Order." March 28, 2017. https://www.youtube.com/watch?v=QevaoEqlgKU.

ABC News. 2019. "President Trump Delivers Remarks on Energy at Shell Pennsylvania Petrochemical Complex." August 12., 2019 https://www.youtube.com/watch?v=zCo9wfcTKl4.

Achenbach, Joel. 2019. "A Remote Virginia Valley Has Been Flooded by Prescription Opioids." *Washington Post*, July 18, 2019. https://www.washingtonpost.com/national/a-remote-virginia -valley-has-been-flooded-by-prescription-opioids/2019/07/18/387bb074-a8ca-11e9-9214 -246e594de5d5_story.html.

Allen, Frederick Lewis. 1954. "Crisis in the Suburbs." *Harper's Magazine* 209 (1250): 47–55.

American Chemistry Council. 2017. "New Report Shows Potential for Major Appalachian Petrochemical Industry." May 18, 2017. https://www.americanchemistry.com/chemistry-in -america/news-trends/press-release/2017/new-report-shows-potential-for-major-appalachian -petrochemical-industry.

Americans for Prosperity. nd. Hot Air Tour. https://www.youtube.com/watch?v=kUofuPGGeQo.

American Petroleum Institute. 2022. "Join Trades Letter to Pelosi and McCarthy." August 11, 2022. https://www.api.org/-/media/Files/News/2022/08/11/Joint-Trades-Letter-Pelosi-McCarthy -IRA-081122.pdf.

Anahkwet. 2023. Interview with Author. June 18, 2023.

Anonymous. 1844. "Are We Free Men?" *Working Man's Advocate*, July 6.

Anonymous. 1865. "The Great West." *The Liberator*, December 29.

Anonymous. 1850. "The World is Full of Beauty." *North Star*, July 11, 1850, p. 4. https://www.loc .gov/resource/sn84026365/1850-07-11/ed-1/?dl=issue&sp=1.

Antonelli, Angela, and Brett Shafer. 1997. "The Road to Kyoto: How the Global Climate Treaty Fosters Economic Impoverishment and Endangers U.S. Sovereignty." Heritage Foundation

Backgrounder, October 6, 1997. https://www.heritage.org/report/the-road-kyoto-how-the-global -climate-treaty-fosterseconomic-impoverishment-and-endangers-us.

Arnade, Chris. 2016. "Pride and Pain in Trump Country." *The Guardian*, September 7, 2016. https:// www.theguardian.com/society/2016/sep/07/kentucky-trump-obama-unemployment-drugs.

Arnold, Ron. 1993. *Ecology Wars: Environmentalism as If People Mattered*. Bellevue, WA: Free Enterprise Press.

Arnold, Ron, and Alan Gottlieb. 1988. *The Wise Use Agenda*. Bellevue, WA: Merril Press.

Auletta, Ken. 1981. "I—The Underclass." *New Yorker*, November 16, 1981. https://www .newyorker.com/magazine/1981/11/16/i-the-underclass.

Badger, Evelyn. 1992. Testimony on the Northern Spotted Owl Preservation Act: Hearing before the subcommittee on Environmental Protection of the Committee on Environment and Public Works, U.S. Senate, 102nd Cong. 48.

Bagdikian, Ben. 1966. "The Rape of the Land." *Saturday Evening Post*, June 18, 1966.

Balzar, John. 1988. "Bush Vows Zero Tolerance of Environmental Polluters." *LA Times*, September 1, 1988, sec. A. https://www.latimes.com/archives/la-xpm-1988-09-01-mn-4551-story.html.

Banking on Climate Chaos. 2021. "Fossil Fuel Finance Report." https://www.bankingonclimat echaos.org/

Banks, Jim, and Kevin McCarthy. 2021. "Banks Working-Class Memo." March 30, 2021. https:// www.documentcloud.org/documents/20534328-banks-working-class-memo.

Bartley, Robert. 1979. "Business and the New Class." In *The New Class?*, edited by B. Bruce Briggs, 57–66. New Brunswick, NJ: Transaction Books.

Bazelon, David. 1979. "How Now 'The New Class'?" *Dissent* (Fall): 443–449.

Beckett, Lois. 2017. "Is There a Neo-Nazi Storm Brewing in Trump Country?" *The Guardian*, June 4, 2017. https://www.theguardian.com/world/2017/jun/04/national-socialism-neo-nazis -america-donald-trump.

Bemiss, Fitzgerald. nd. "The Future of Public Education in Virginia." Virginia Museum of History and Culture, Fitzgerald Bemiss Papers, 1952–1988, Collection Number Mss1 B4252 a FA2.

Bemiss, Fitzgerald. nd. "Letter to James Buchanan." Virginia Museum of History and Culture, Fitzgerald Bemiss Papers, 1952–1988, Collection Number Mss1 B4252 a FA2.

Bemiss, Fitzgerald. 1955. "Letter to Prescott S. Bush." December 21, 1955. University of Virginia Library, Papers of James S. Kilpatrick. Collection Number 6626-b. Charlottesville, Virginia.

Bemiss, Fitzgerald. 1960. "Address to the Junior Chamber of Commerce." March 16, 1960. Virginia Museum of History and Culture, Fitzgerald Bemiss Papers, 1952–1988, Collection Number Mss1 B4252 a FA2. Richmond, VA.

Bemiss, Fitzgerald. 1964. "Address to the 15th Annual Convention–Virginia State Division of IWLA." Virginia Museum of History and Culture, Fitzgerald Bemiss Papers, 1952–1988, Collection Number Mss1 B4252 a FA2. Richmond, VA.

Bemiss, Fitzgerald. 1970. "Virginia Citizens Planning Commission." May 4. Virginia Outdoor Recreation Commission folder. Virginia Museum of History and Culture, Fitzgerald Bemiss Papers, 1952–1988, Collection Number Mss1 B4252 a FA2. Richmond, VA.

Bemiss, Fitzgerald, Harry Byrd Jr., A. Plunket Beirne, James Camblos, Walther Fidler, Ira Gabrielson, Cecil Gilkerson, Meriweather Lewis, Paul Manns, Floyd McKenna, George McMath, Dorman Miller, Meade Palmer, Stockton Tyler, and Conrad Wirth. 1965. "Virginia's Common Wealth: A Study of Virginia's Outdoor Recreation Resources." Virginia Outdoor Recreation Study Committee, November 1, 1965.

Benham, Steve. 2020. "Oregon Democrats Issue Subpoenas to Absent GOP House Members, Courtney Pleads for Return." KATU, February 7, 2020. https://katu.com/news/politics/oregon-democrats-issue-subpoenas-to-absent-gop-house-members-courtney-pleads-for-return.

Berry, Jonathan. 2021. "The First Week of the Biden Administration." *First Things*, January 29, 2021. https://www.firstthings.com/web-exclusives/2021/01/the-first-week-of-the-biden-administration.

Beutler, Brian. 2009. "AFL-CIO Urges Congressmen to Pass Climate Bill." *Talking Points Memo*, June 25, 2009. https://talkingpointsmemo.com/dc/afl-cio-urges-congressmen-to-pass-climate-bill.

Brown, Matthew, and Michael Phillis. 2022. "Climate Change? The Inflation Reduction Act's Surprise Winner, the US Oil and Gas Industry." *USA Today*, August 18, 2022.

Bruce-Briggs, B. 1979. *The New Class?* New York: McGraw-Hill.

Bruggers, James, and Joseph Gerth. 2016. "Paul, McConnell Rip Clinton Coal Jobs Remark." *Courier-Journal*, March 14, 2016. https://www.alamogordonews.com/story/news/politics/2016/03/14/rand-paul-rips-clintons-coal-jobs-remark/81772992/.

Buchanan, Pat. 1976. "Why the Right Failed." *Conservative Digest* 2:15–16.

Buchanan, Pat. 2020. "In the Pandemic, It's Every Nation for Itself." *Muskogee Phoenix*, March 18, 2020. https://www.muskogeephoenix.com/opinion/columns/buchanan-in-the-pandemic-its-every-nation-for-itself/article_881da21a-661a-5d4d-8c9a-e359dce0d8fa.html.

Bureau of Labor Statistics. 2023. "Work Stoppages Summary." February 22, 2023. https://www.bls.gov/news.release/wkstp.nr0.htm.

Burie, Dale. 2018. Interview with Author. March 8, 2018.

Burton, Hal. 1955. "Trouble in the Suburbs." *Saturday Evening Post*, September 17, 1955, pp. 19–21, 113–118.

Byrd-Hagel Resolution. 1997. Senate Foreign Relations Committee. 105th Congress. https://www.congress.gov/bill/105th-congress/senate-resolution/98/text.

Cannon, Lou. 1973. "Ecology Drive Draws Near Collision with Fiscal Barrier." *Washington Post*, February 14, 1973.

ChildCare Aware of America. 2021. "Catalyzing Growth: Using Data to Change Child Care." https://www.childcareaware.org/catalyzing-growth-using-data-to-change-child-care/#ChildCare Affordability.

City Care. 1979. "City Care: A National Conference on the Urban Environment." April 8–11, 1979. Environmentalists for Full Employment Records, 1969–1984, AIS.1984.27. Archives and Special Collections, University of Pittsburgh.

Clean Air Council. 2014. "Health Impact Assessment of the Shell Chemical Appalachia Petrochemical Complex." https://cleanair.org/wp-content/uploads/HIA-Final.pdf.

Climate Justice Alliance. 2020. "Press Release: Climate Justice Alliance Demands an Immediate End to Trump's Xenophobia and Negligence." March 12, 2020. https://climatejusticealliance.org /climate-justice-alliance-demands-an-immediate-end-to-trumps-xenophobia-and-negligence/.

Clinton, Bill. 1992. "A New Covenant for Environmental Progress." April 22, 1992. https://www .c-span.org/video/?25724-1/clinton-campaign-speech.

Cohn, Nate. 2016. "Why Trump Won: Working-Class Whites." *New York Times*, November 9, 2016. https://www.nytimes.com/2016/11/10/upshot/why-trump-won-working-class-whites.html.

Collins, Chuck, and Omar Ocampo. 2021. "Trump and His Many Billionaire Enablers." Inequality .org, January 11, 2021. https://inequality.org/great-divide/trump-many-billionaire-enablers/.

Colvin, Jill. 2016. "Trump Is Hammering Clinton on Coal." *Business Insider*, May 6, 2016. https:// www.businessinsider.com/ap-trump-returns-to-campaign-trail-targets-clinton-on-coal-2016-5.

Coptis, Veronica. 2021. Interview with Author. August 11, 2021.

CNN. 2016. "Transcript of Democratic Town Hall at Ohio State University." March 13, 2016. https://transcripts.cnn.com/show/se/date/2016-03-13/segment/02.

Coyne, John. 1972. *The Impudent Snobs: Agnew vs. the Intellectual Establishment*. New Rochelle, NY: Arlington House.

Crane, Philip. 1978. "The Blue Collar Constituency Is Really Conservative." *Conservative Digest* 4:6.

Crittenden, Ann. 1982. "A Stubborn Chamber of Commerce Roils the Waters." *New York Times*, June 27, 1982. https://www.nytimes.com/1982/06/27/business/a-stubborn-chamber-of -commerce-roils-the-waters.html.

C-SPAN. 2016. "Road to the White House 2016." May 5, 2016. https://www.c-span.org/video /?409094-1/presidential-candidate-donald-trump-rally-charleston-west-virginia.

Davis, Bob, and Rebecca Ballhaus. 2016. "The Place That Wants Donald Trump the Most." *Wall Street Journal*, April 17, 2016. https://www.wsj.com/articles/the-place-that-loves-donald-trump -most-1460917663.

Davis, Rob. 2019. "Polluted by Money." *The Oregonian*, February 22, 2019. https://projects .oregonlive.com/polluted-by-money/part-1.

Democracy Now. 2020. "A Class Rebellion: Keeanga-Yamahtta Taylor on How Racism & Racial Terrorism Fueled Nationwide Anger." June 1, 2020. https://www.democracynow.org/2020/6/1 /keeanga_yamahtta_taylor_protests_class_rebellion.

Democratic Leadership Council. 1990. "The New Orleans Declaration: A Democratic Agenda for the 1990s." Fourth Annual Conference of the Democratic Leadership Council, New Orleans, LA, March 22–25, 1990.

Derman, Joanna. 2022. "The Inflation Reduction Act: Topline Oil and Gas Reforms." Project on Government Oversight, September 2, 2022. https://www.pogo.org/analysis/2022/09/the-inflation -reduction-act-topline-oil-and-gas-reforms.

De Rugy, Veronique. 2022. "Dems Taking from Working-Class, Giving to Laptop Class." *Boston Herald*, September 5, 2022. https://www.bostonherald.com/2022/09/05/de-rugy-dems-taking -from-working-class-giving-to-laptop-class/.

DiChristopher, Tim. 2019. "Ocasio-Cortez's Green New Deal Is Not Going Over Well at One of the Year's Biggest Energy Gatherings." CNBC, March 12, 2019. https://www.cnbc.com/2019/03 /12/aocs-green-new-deal-not-going-over-well-at-ceraweek-energy-conference.html.

Dickinson, Tim. 2019. "Runaway Senators, Militias and Koch Money: What the Hell Just Happened in Oregon?" *Rolling Stone*, June 27, 2019. https://www.rollingstone.com/politics/politics -news/oregon-climate-battle-gop-walkout-violence-852643/.

Dieter, Cheryl, Molly Maupin, Rodney Caldwell, Melissa Harris, Tamara Ivahnenko, John Lovelace, Nancy Barber, and Kristin Linsey. 2018. "Estimated Use of Water in the United States in 2015." United States Geological Survey Circular 1441. https://doi.org/10.3133/cir1441

Dobbs, Kevin. 2022. "Why Is Inflation Reduction Act a Big Deal for Natural Gas, Oil Industries?" *Natural Gas Intelligence*, August 8, 2022. https://www.naturalgasintel.com/why-is-inflation -reduction-act-a-big-deal-for-natural-gas-oil-industries/.

Duffy, Clare, and Ross Levitt. 2020. "Energy Companies Cancel Construction of Atlantic Coast Pipeline." CNN, July 5, 2020. https://www.cnn.com/2020/07/05/us/duke-dominion-energy -cancel-atlantic-coast-pipeline/index.html.

Earth First! 2018. "'The Fire Is Catching': Mountain Valley Pipeline Faces Fierce Opposition in the Virginias." *Earth First Journal*, May 18, 2018. https://earthfirstjournal.org/newswire/2018/05/18 /the-fire-is-catching-mountain-valley-pipeline-faces-fierce-opposition-in-the-virginias/.

EarthJustice. 2021. "Back Forty Mine Developer Relinquishes Mining Permits amid Legal Setbacks." May 11, 2021, https://earthjustice.org/press/2021/back-forty-mine-developer-relinquishes -mining-permits-amid-legal-setbacks.

Eastland, James. 1955. Speech on Senate Floor. May 26, 1955. Congressional record, 84th Cong., 1st Sess, 7124.

Eberhart, Dan. 2022. "The Inflation Reduction Act Is No Oil Killer." *Forbes*, August 12, 2022. forbes.com/sites/daneberhart/2022/08/12/the-inflation-reduction-act-is-no-oil-killer/?sh =b0e620e2905e.

El Dahan, Maha, and Jan Strupczewski. 2023. "Davos 2023: EU to Counter U.S. Climate Game Changer with Own Green Deal." Reuters, January 17, 2023. https://www.reuters.com/business /sustainable-business/davos-iea-director-says-inflation-reduction-act-is-new-climate-accord-2023 -01-17/.

Ellison, Garret. 2022. "In the UP, a New Chapter Begins in 20-Year Clash over Gold Mine." *MLive*, August 4, 2022. https://www.mlive.com/public-interest/2022/08/in-the-up-a-new-chapter-begins -in-20-year-clash-over-gold-mine.html.

Engelfried, Nick. 2020. "Indigenous-Led Resistance to Enbridge's Line 3 Pipeline Threatens Big Oil's Last Stand." Waging Nonviolence, December 14, 2020. https://wagingnonviolence.org/2020/12/indigenous-water-protectors-enbridge-line-3-pipeline-big-oil-last-stand/.

Engel, John. 2018. "Coalition to Save the Menominee River." March 2018.http://savethewildup.org/2018/03/coalition-to-save-the-menominee-river/.

EFFE (Environmentalists for Full Employment). 1980. "Summary of [Safe Energy and Full Employment] Conference." Environmentalists for Full Employment Records, 1969–1984, AIS.1984.27, Archives & Special Collections, University of Pittsburgh, Box 9, Folder 5.

EFFE (Environmentalists for Full Employment). 1981. "No More Three Mile Islands!" Environmentalists for Full Employment Records, 1969–1984, AIS.1984.27, Archives & Special Collections, University of Pittsburgh, Box 4, folder 116.

Evans, George Henry. 1844. "Equal Rights to Land." *Working Man's Advocate*, March 16, 1844.

Everett, Burgess, and Marianne Levine. 2022. "Manchin's Latest Shocker: A 700B Deal." *Politico*, July 27, 2022. https://www.politico.com/news/2022/07/27/manchin-schumer-senate-deal-energy-taxes-00048325.

Ewing, Jack. 2023. "Battle Over Electric Vehicles Is Central to Auto Strike." *New York Times*, September 16, 2023. https://www.nytimes.com/2023/09/16/business/electric-vehicles-uaw-gm-ford-stellantis.html.

Falders, Katherine, and Candace Smith. 2016. "Trump Caters to Coal Country in West Virginia Rally." *ABC News*, May 5, 2016.

Finn, Teaganne, and Julie Tsirkin. 2021. "Manchin Says He's a 'No' on Biden's Build Back Better Legislation." *NBC News*, December 19, 2021. https://www.nbcnews.com/politics/congress/manchin-says-he-no-biden-s-build-back-better-legislation-n1286281.

Fisher, Marc. 2016. "In West Virginia, Coal Country Voters Are Thrilled about Donald Trump." *Washington Post*, December 6, 2016. https://www.washingtonpost.com/politics/in-west-virginia-coal-country-voters-are-thrilled-about-donald-trump/2016/12/06/8eb0b0ca-b8c2-11e6-b994-f45a208f7a73_story.html.

Frazier, Reid. 2022. "Families Are Leaving Beaver County as Shell's Ethane Cracker Sets to Open." WESA, July 31, 2022. https://www.wesa.fm/environment-energy/2022-07-31/families-leave-beaver-shell-ethane-cracker.

Frink, Henry, and Warren Butler. 1844. "Working Man's Movement in Otsego Co.: The Ball in Motion." *Working Man's Advocate*, July 6, 1844.

Funk, Ashley. 2021. Interview with Author. September 30.

Furgurson, Ernest. 1970. "The Black Version of Earth Day." *Baltimore Sun*, April 14, 1970. https://web.sas.upenn.edu/earthdayproject/civil-rights/

Galston, William, and Elaine Kamarck. 1989. *The Politics of Evasion: Democrats and the Presidency.* Washington, DC: Progressive Policy Institute.

GCC (Global Climate Coalition). nd. "Economic and Employment Impacts from Proposed Greenhouse Gas Restrictions." https://www.industrydocuments.ucsf.edu/fossilfuel/docs/#id=fqfl0228

GCC (Global Climate Coalition). 1997a. "Americans Work Hard for What We Have, Mr. President." *New York Times*, June 19, 1997. https://www.climatefiles.com/denial-groups/global-climate-coalition-collection/1997-anti-kyoto-ads/.

GCC (Global Climate Coalition). 1997b. "The Only Thing This Treaty Cools Down." *Washington Post*, November 13, 1997, A20. https://embed.documentcloud.org/documents/7275864-Kyoto-Opposition-Washington-Post-Ads-1997.

GCC (Global Climate Coalition). 1997c. "US Chamber President Challenges White House Climate Change Policies." Press Release, October 5, 1997. Chamber of Commerce of the United States Records (1960), Series IV, Box 67. Hagley Museum & Library, Wilmington, DE.

GCC (Global Climate Coalition). 1998. "Labor Voices on Kyoto." https://www.industrydocuments.ucsf.edu/fossilfuel/docs/#id=pfgl0228

GCC (Global Climate Coalition). 2001. Home page. https://web.archive.org/web/20010302000601/http://www.globalclimate.org/index.htm.

Gedicks, Al. 2023. Interview with author. June 9, 2023.

Gemen, Ben. 2009. "Can Push for Climate Bill Forge a Lasting Labor-Enviro Alliance?" *New York Times* ClimateWire, August 28, 2009. https://archive.nytimes.com/www.nytimes.com/cwire/2009/08/28/28climatewire-can-push-for-climate-bill-forge-a-lasting-la-70854.html?pagewanted=all.

Gilder, George. 1993 [1981]. *Wealth & Poverty*. Oakland: ICS Press.

Grossman, Richard. 1976. Letter to Representative John Conyers. August 11, 1976. Environmentalists for Full Employment Records, 1969–1984, AIS.1984.27, Archives & Special Collections, University of Pittsburgh, Box 6, Folder 166.

Guo, Jeff. 2016. "A New Theory for Why Trump Voters are So Angry - That Actually Makes Sense." *Washington Post*, November 8, 2016. https://www.washingtonpost.com/news/wonk/wp/2016/11/08/a-new-theory-for-why-trump-voters-are-so-angry-that-actually-makes-sense/

Handler, Brad, and Morgan Bazilian. 2022. "The Inflation Reduction Act's Modest Impact on Oil and Gas." *The Hill*, August 22, 2022. https://thehill.com/opinion/energy-environment/3611212-the-inflation-reduction-acts-modest-impact-on-oil-and-gas/.

Hart, Jeffrey. 1975. "Emerging 'New Class' Splits Democrats." *Danville Register*, November 9, 1975.

Hass, Patricia. 2013. *Monument Avenue Memories: Growing Up on Richmond's Grand Avenue*. Charleston, SC: The History Press.

Hernandez, Ernie. 1981. "Nuclear Meet Equates N-Power, N -Bombs." *Gary Post Tribune*, November 22, 1981. Environmentalists for Full Employment Records, 1969–1984, AIS.1984.27, Archives and Special Collections, University of Pittsburgh, Box 2, Folder 32.

Hersh, Adam. 2023. "UAW-Automakers Negotiations Pit Falling Wages against Skyrocketing CEO Pay." Economic Policy Institute (blog). September 12, 2023. https://www.epi.org/blog/uaw-automakers-negotiations/

Hiltzik, Michael. 2019. "Shell Oil Allegedly Coerced Its Workers to Attend a Trump Rally. Blame the Supreme Court." *LA Times*, August 19, 2019. https://www.latimes.com/business/story/2019 -08-19/shell-oil-trump-rally.

Hornblower, Margot. 1979. "Major Industries Map New Attack on Clean Air Act." *Washington Post*, January 15, 1979. https://www.washingtonpost.com/archive/politics/1979/01/15/major -industries-map-new-attack-on-clean-air-act/197a03d9-3225-4488-8a85-7ce474681805/.

IPCC (Intergovernmental Panel on Climate Change). 2022. Summary for Policymakers. In Climate Change 2022: Impacts, Adaptation, and Vulnerability. https://www.ipcc.ch/report/ar6/wg2 /downloads/report/IPCC_AR6_WGII_SummaryForPolicymakers.pdf.

Ireland, Karan. 2021. Interview with author. August 2, 2021.

Ivins, Molly. 1979. "Dam Break Investigated: Radiation of Spill Easing." *New York Times*, July 28, 1979. https://timesmachine.nytimes.com/timesmachine/1979/07/28/111730084.html ?pageNumber=6.

Jacobs, Ben. 2016. "Confident Donald Trump Tells His West Virginia Supporters: Don't Bother Voting." *The Guardian*, May 5, 2016. https://www.theguardian.com/us-news/2016/may/05 /donald-trump-west-virginia-campaign-rally-election-2016.

Jacobs, Jeremy. 2016. "Why the Environment Wasn't a Factor in the Presidential Race." *E&E News*, November 22, 2016. https://www.eenews.net/stories/1060046166.

Kahn, Brian, and Dhruv Mehrotra. 2021. "The Big Oil Money behind the Members of Congress Who Fueled the Capitol Attack." *Gizmodo*, January 8, 2021. https://gizmodo.com/the-big-oil -money-behind-the-members-of-congress-who-fu-1846020442.

Kalhoefer, Kevin. 2016. "Primary Debate Scorecard: Climate Change through 20 Presidential Debates." *Media Matters*, March 23, 2016. https://www.mediamatters.org/donald-trump/primary -debate-scorecard-climate-change-through-20-presidential-debates.

Kaplan, Thomas. 2016. "This Is Trump Country." *New York Times*, March 4, 2016. https://www .nytimes.com/interactive/2016/03/04/us/politics/donald-trump-voters.html.

KATU. 2019. "Local Timber Industry Workers Rally in Salem to Protest Cap and Trade Bill." June 19, 2019. https://katu.com/news/local/timber-industry-workers-rally-in-salem-to-protest -cap-and-trade-bill

KATU. 2021. "Timber Unity Responds to Rumors it Was at the Riot and Rally in D.C." January 7, 2021. https://ktvl.com/news/local/timber-unity-responds-to-it-was-at-the-rally-and -riot-in-dc

Kaus, Mickey. 1990. *The Commercial Appeal*. (Memphis, TN), September 21, 1990, A10.

Kemp, Jack. 1982. "Wide Support for the Conservative Revolution." *Conservative Digest*, October 1982, 8–9.

Kieschnick, Michael. 1978. *Environmental Protection and Economic Development*. Washington, DC: US Department of Commerce Economic Development Administration, October, 26

Kirong, Nephele. 2022. "NextEra Takes $800M Impairment Charge on Mountain Valley Pipeline Investment, S&P Global Market Intelligence." February 21, 2022. https://www.spglobal.com /marketintelligence/en/news-insights/latest-news-headlines/nextera-takes-800m-impairment -charge-on-mountain-valley-pipeline-investment-69002279.

Kormann, Carolyn. 2019. "How Rogue Republicans Killed Oregon's Climate Change Bill." *New Yorker*, June 28, 2019. https://www.newyorker.com/news/news-desk/how-rogue-republicans -killed-oregons-climate-change-bill.

Kristol, Irving. 1975. "Business and 'the New Class.'" *Wall Street Journal*, May 19, 1975.

Kroll, Andy. 2021. "What Insurrection: Corporate America Can't Stop Bankrolling the Jan. 6 Sedition Caucus." *Rolling Stone*, October 25, 2021. https://www.rollingstone.com/politics/politics -features/corporate-donors-sedition-caucus-insurrection-1247707/.

Krupp, Frederic. 1986. "New Environmentalism Factors in Economic Needs." *Wall Street Journal*, November 20, 1986. https://www.wsj.com/articles/SB117269353475022375.

Kuttner, Robert. 1984. "Jobs." *Dissent*, Winter 1984. https://www.dissentmagazine.org/article /jobs/

Labor Network for Sustainability. 2016. "Just Transition, Just What is It?" https://www .labor4sustainability.org/files/Just_Transition_Just_What_Is_It.pdf.

Labor Network for Sustainability. 2023. "An Open Letter to Big 3 Auto CEOs." https://www .labor4sustainability.org/uaw-solidarity-letter/.

Ladd, Everett Carll. 1979. "Pursuing the New Class: Social Theory and Survey Data." In *The New Class?*, edited by B. Bruce-Briggs. New York: McGraw Hill.

Larsen, Brooke. 2024. "Labor Unions and Environmentalists Are Working Together on the Energy Transition." *High Country News*, January 1, 2024. https://www.hcn.org/issues/56-1/climate -change-labor-unions-and-environmentalists-are-working-together-on-the-energy-transition/.

Leber, Rebecca, and Ali Breland. 2020. "The Oregon GOP's Favorite Anti-Environment Group Is Awash in Racism and Violent Threats." *Mother Jones*, March 6, 2020. https://www.motherjones .com/politics/2020/03/timber-unity-racism/.

Lewis, Paul, Tom Silverstone, and Adithya Sambamurthy. 2016. "Why the Poorest County in West Virginia Has Faith in Trump." *The Guardian*, October 13, 2016. https://www.theguardian.com/us -news/video/2016/oct/12/west-virginia-donald-trump-supporters-mcdowell-county-poverty-video.

Lindwall, Courtney, and Melissa Denchak. 2022. "What Is the Keystone XL Pipeline." Natural Resources Defense Council. March 15, 2022. https://www.nrdc.org/stories/what-keystone-xl -pipeline#whatis.

Litvak, Anya. 2019. "Trump's Large Union Crowd at Shell Was Given the Option of Not Showing Up—and Not Getting Paid." *Pittsburgh Post-Gazette*, August 16, 2019.

Lowry, Rich. 2020. "We Are All Restrictionists Now." *Politico*, April 1, 2020. https://www.politico .com/news/magazine/2020/04/01/we-are-all-restrictionists-now-160459.

MacFarquhar, Larisa. 2016. "In the Heart of Trump Country." *New Yorker*, October 16, 2016. https://www.newyorker.com/magazine/2016/10/10/in-the-heart-of-trump-country.

Macrotrends. 2023. "Oil Price History Chart." https://www.macrotrends.net/1369/crude-oil-price-history-chart.

Manchin, Joe. "Dead Aim," October 9, 2010. https://www.youtube.com/watch?v=xIJORBR-pOPM.

Maraniss, David, and Michael Weisskopf. 1987. "Jobs and Illness in Petrochemical Corridor." *Washington Post*, December 22, 1987.

Mazurek, Jan. 2003. "Cap Carbon Dioxide Now." PPI Front and Center, January 7, 2003. http://www.ndol.org/ndol_sub_kaid_116_subid_149.html.

McCammon, Sarah. 2015. "Trump Draws Large Crowd, Attacks Rivals in South Carolina Campaign Stop." NPR, July 22, 2015. https://www.npr.org/2015/07/22/425225015/trump-draws-large-crowd-attacks-rivals-in-south-carolina-campaign-stop.

McConnell, Mitch. 2022. "Democrats' Inflation Bill Will Cut Jobs and Wages but Not Inflation." YouTube video, August 2, 2022. https://www.youtube.com/watch?v=SXzY1JHgeNU.

McGinty, Kathleen. 2004. "Environmental Armistice." *DLC Blueprint Magazine*, March 23, 2004. http://www.ndol.org/ndol_ci_kaid_116_subid_150_contentid_252477.html.

McLelland, Edward. 2016. "The Rust Belt Was Turning Red Already." *Washington Post*, November 9, 2016. https://www.washingtonpost.com/posteverything/wp/2016/11/09/the-rust-belt-was-turning-red-already-donald-trump-just-pushed-it-along/.

Media Matters Staff. 2022. "Charlie Kirk Likens Concern about Climate Change to 'Pseudo Paganism." Media Matters for America, August 8, 2022. https://www.mediamatters.org/charlie-kirk/charlie-kirk-likens-concern-about-climate-change-pseudo-paganism.

Meese, William. 1973. "The Man-Made Energy Crisis." NAM Reports, September 3, 1973. National Association of Manufacturers records, Acc. 1411, Box 677, Environmental Quality and Conservation Department.

Milman, Oliver. 2021. "A Closer Look at Joe Manchin's Ties to the Fossil Fuel Industry." *Mother Jones*, October 21, 2021. https://www.motherjones.com/politics/2021/10/a-closer-look-at-joe-manchins-ties-to-the-fossil-fuel-industy/.

Mitchell, Maurice. 2022. "Building Resilient Organizations." *The Forge*, November 29, 2022. https://forgeorganizing.org/article/building-resilient-organizations.

Murray, Charles. 1990. "The Underclass, Revisited." American Enterprise Institute, January 1, 1990. https://www.aei.org/research-products/working-paper/the-underclass-revisited/.

Murray, Charles. 1994 [1984]. *Losing Ground: American Social Policy: 1950–1980*. New York: Basic Books.

National Alliance to End Homelessness. 2023. "State of Homelessness." https://endhomelessness.org/homelessness-in-america/homelessness-statistics/state-of-homelessness/

National Association of Manufacturers. 1973a. "NAM Regrets Price Freeze." *NAM Reports* 18 (June 18): 1.

National Association of Manufacturers. 1973b. "Industry Stresses Cooperation on Pollution Curbs." *NAM Reports* 18 (June 18): 1.

New York Post Editorial Board. 2016. "Hillary's Vow to Kill Coal Miners' Jobs Finishes a Vast Democratic Betrayal." *New York Post*, March 16, 2016. https://nypost.com/2016/03/14/hillarys -vow-to-kill-coal-miners-jobs-finishes-a-vast-democratic-betrayal/.

New York Times. 2023. "A Partial List of U.S. Mass Shootings in 2023." June 21, 2023. https://www .nytimes.com/article/mass-shootings-2023.html.

Nobel, Justin. 2020. "Whose Allegiance? Three Percenters Militia Working in Bakken Oil Patch Raises Concerns of Domestic Terrorism Risk." *DeSmog News*, July 21, 2020.

Novak, Michael. 1973. "The Center and the Left." *Commonweal*, January 12, 1973, pp. 318, 335.

Nunes, Devin. 2009. "Top Secret Democrat Plan Exposed: Cap and Tax." YouTube video. https:// www.youtube.com/watch?v=aGLeya3_Bws.

NY Times editors. 1986. "A Bhopal on the Bayou?" *New York Times*, January 5, 1986, p. 15.

Oil, Chemical and Atomic Workers International Union. 1976. "Health, Safety and Environmental Considerations as Factors in OCAW Plant Closings, 1970–75." Denver: OCAW Research Department.

O'Leary, Sean. 2021. "The Natural Gas Fracking Boom and Appalachia's Lost Economic Decade." Ohio River Valley Institute, February 12, 2021. https://ohiorivervalleyinstitute.org/new-report -natural-gas-county-economies-suffered-as-production-boomed/.

Overton, Justinn. 2021. Interview with author. August 4, 2021.

Paine, Barbara. 1958. "Preservation of Open Spaces offers a Challenge to Suburbia." *Nature* 51: 428–430.

Payton, Bre. 2016. "Hillary Clinton Has a Message for Coal Miners: You're Fired." *The Federalist*, March 14, 2016. https://thefederalist.com/2016/03/14/hillary-clinton-has-a-message-for-coal -miners-youre-fired/.

Perisco, Emily, Rob Alternburg, and Christina Simeone. 2021. "Buried Out of Sight: Uncovering Pennsylvania's Hidden Fossil Fuel Subsidies." PennFuture. https://www.pennfuture.org/Files /Admin/PF_FossilFuel_Report_final_2.12.21.pdf.

Peters, Charles. 1982. "A Neo-Liberal's Manifesto." *Washington Post*, September 5, 1982. https:// www.washingtonpost.com/archive/opinions/1982/09/05/a-neo-liberals-manifesto/21cf41ca-e60e -404e-9a66-124592c9f70d/.

Peters, Charles, and Phillip Keisling. 1985. *A New Road for America: The Neoliberal Movement*. Lanham, MD: Madison Books.

Pilkington, Ed, and Mona Chalabi. 2016. "Climate Change: The Missing Issue of the 2016 Campaign." *The Guardian*, July 5, 2016. https://www.theguardian.com/us-news/2016/jul/05/climate -change-voters-2016-election-issues.

Podhoretz, Norman. 1979. "The Adversary Culture and the New Class." In *The New Class?*, edited by B. Bruce-Briggs. New York: McGraw Hill.

Polumbo, Brad. 2022. "The Democrats' 'Inflation Reduction Act' Betrays the People Progressives Claim to Care About." *Newsweek*, August 16, 2022. https://www.newsweek.com/democrats -inflation-reduction-act-betrays-people-progressives-claim-care-about-opinion-1734247.

Powell, Lewis. 1971. "Attack on American Free Enterprise System." April 23, 1971. https:// scholarlycommons.law.wlu.edu/powellmemo/1/

Profita, Cassandra. 2020. "With Oregon Climate Action Stalled in Salem, Some Businesses Look beyond Cap and Trade." Oregon Public Broadcasting, March 4, 2020. https://www.opb.org/news /article/oregon-cap-trade-bill-legislation-climate-change-businesses/.

Rees, Colin. 2022. "People vs. Fossil Fuels Coalition Responds to Inflation Reduction Act." *Oil Change International*, August 12, 2022. https://priceofoil.org/2022/08/12/pvff-inflation-reduction -act-manchin-side-deal/.

Republican National Committee. 2016. Party Platform. https://www.presidency.ucsb.edu/sites /default/files/books/presidential-documents-archive-guidebook/national-political-party-platforms -of-parties-receiving-electoral-votes-1840-2016/117718.pdf.

Rice, John. 1979. "Sagebrush Rebellion Erupts in Both Houses." *Reno Gazette Journal*, February 15, 1979.

Ring, Edward. 2020. "Twin Paths to Socialism: 'Equity' and 'Climate Change' Alarmism." *American Greatness*, November 15, 2020. https://amgreatness.com/2020/11/15/the-twin-paths-to -socialism-equity-and-climate-change-alarmism/.

River Alliance of Wisconsin. 2018. "Back 40 Mine Update and Action Alert." May 10, 2018. https://www.wisconsinrivers.org/back-40-contact-investors/.

Roberts, David. 2020. "Oregon Republicans Are Subverting Democracy by Running Away Again." *Vox*, February 29, 2020. https://www.vox.com/energy-and-environment/2020/2/29/21157246 /oregon-republicans-walk-out-climate-change-cap-trade-democracy.

Robertson, Gary. 2014. "Power, Politics, and Powell." *Richmond Magazine*, July 14, 2014. https:// richmondmagazine.com/news/features/lewis-powell-jr-manifesto/.

Rothenberg, Stuart. 2019. "Why Working-Class Whites Aren't Giving Up on Trump." *Roll Call*, September 10, 2019. https://rollcall.com/2019/09/10/why-working-class-whites-arent-giving-up -on-trump/.

Rowland, Stanley. 1957. "Flight to Suburbia." *The Nation* 184: 38–39.

Ruffini, Patrick. 2023. "The Emerging Working-Class Republican Majority." *Politico*, November 4, 2023. https://www.politico.com/news/magazine/2023/11/04/new-republican-party-working -class-coalition-00122822.

Rysavy, Tracy Fernandez. 2017. "The Back Forty Mine: Is It the Next Standing Rock?" *Green America*, Fall. https://www.greenamerica.org/drinking-water-risk/back-forty-mine-it-next-standing -rock.

Sanders, Bernie. 2020. "The Green New Deal." https://int.nyt.com/data/documenthelper/1654
-bernie-sanders-green-new-deal/761873c26ec4075c609b/optimized/full.pdf#page=1.

Saward, John. 2016. "Welcome to Trump County, USA." *Vanity Fair*, February 24, 2016. https://
www.vanityfair.com/news/2016/02/donald-trump-supporters-west-virginia.

Schapiro, Jeff. 2011. "Civic Leader, Conservation Advocate Fitzgerald Bemiss Dies at 88." *Richmond
Times-Dispatch*, February 8, 2011. https://richmond.com/entertainment/civic-leader-conservation
-advocate-fitzgerald-bemiss-dies-at-88/article_f3328db5-61aa-5a52-80d0-4aea632c73b4.html.

Schneider, Elena. 2022. "Working-Class Struggles Shake Nevada, Threatening Democratic Party."
Politico, October 26, 2022. https://www.politico.com/news/2022/10/26/democrats-working-class
-nevada-00063465.

Schulte, Laura. 2021. "Michigan Judge Denies Wetland Permit for Back Forty Mine along
Menominee River." *Milwaukee Journal Sentinel*, January 5, 2021. https://www.jsonline.com/story
/news/2021/01/05/michigan-judge-denies-wetland-permit-back-forty-mine-citing-incomplete
-information/4139460001/.

Schwartz, Brian. 2022a. "Koch Network Pressures Sens. Manchin, Sinema to Oppose $739 Bil-
lion Tax-and-Spending Bill." CNBC, August 1, 2022. https://www.cnbc.com/2022/08/01/koch
-network-pressures-sens-manchin-sinema-to-oppose-739-billion-tax-and-spending-bill.html.

Schwartz, Brian. 2022b. "How Wall Street Wooed Sen. Kyrsten Sinema and Preserved Its Multi-
billion Dollar Carried Interest Tax Break." CNBC, August 9, 2022. https://www.cnbc.com/2022
/08/09/how-wall-street-wooed-sen-kyrsten-sinema-and-preserved-its-multi-billion-dollar-carried
-interest-tax-break.html.

Selsky, Andrew. 2020. "Conservative, Rural Groups Rally to Protest Oregon Climate Bill." *The
Columbian*, February 6, 2020. https://www.columbian.com/news/2020/feb/06/conservative-rural
-groups-rally-to-protest-oregon-climate-bill/.

Senn, Milton. 1963. "We Must Stop Contaminating Our Water." *American Home*, January/Feb-
ruary, 1963.

Shabecoff, Philip. 1979. "Big Business on the Offensive." *New York Times*, December 9.

Sierra Club. nd. Ohio. "Petrochemicals in the Ohio River Valley." https://www.sierraclub.org/ohio
/petrochemicals.

Slipek, Edwin Jr. 1980. "Architecture: Sincerest Flattery." *Style Weekly*, January 1, 1980. https://
www.styleweekly.com/richmond/achitecture-sincerest-flattery/Content?oid=1390879.

Smith, Jason. 2022. "The Inflation Reduction Act Will Prolong and Make Worse Biden's Inflation Crisis." *Townhall*, August 19, 2022. https://townhall.com/capitol-voices/repjasonsmith
/2022/08/19/the-inflation-reduction-act-will-prolong-and-make-worse-bidens-inflation-crisis
-n2611974.

Sozzi, Brian. 2022. "'He Doesn't Allow Us to Drill': Home Depot Co-Founder Bernie Marcus
Blames President Biden for Oil Supply Issues." Yahoo Finance, October 18, 2022. https://finance
.yahoo.com/news/home-depot-co-founder-bernie-marcus-president-biden-oil-110259529.html.

Spect, Joshua. 2019. "Hamburgers Have Been Conscripted into Fights over the Green New Deal." *Time*, May 7, 2019. https://time.com/5583986/green-new-deal-beef-history/.

Starr, Roger and James Carlson. 1968. "Pollution and Poverty: The Strategy of Cross-Commitment." *The Public Interest* 10: 104–131.

Stetson, Stephen. 2021. Interview with author. August 2, 2021.

StopLine3.org. nd. "Stop the Line 3 Pipeline." https://www.stopline3.org/#intro.

Taub, Amanda. 2016. "Behind 2016's Turmoil, a Crisis of White Identity." *New York Times*, November 1, 2016. https://www.nytimes.com/2016/11/02/world/americas/brexit-donald-trump-whites.html.

Tavernise, Sabrina. 2021. "Virginia Removes Robert E. Lee Statue from State Capital." *New York Times*, September 8, 2021. https://www.nytimes.com/2021/09/08/us/robert-e-lee-statue-virginia.html.

Taylor, Jared. 2020. "No Borders Crowd Eats Crow." *American Renaissance*, April 3, 2020. https://www.amren.com/podcasts/2020/04/no-borders-crowd-eats-crow/.

The Progress. 1980. "Reagan Supports 'Sagebrush Rebellion.'" *The Progress* (Fillmore, Utah), November 28, 1980, p. 1.

Therrien, Jim. 2018. "Bennington's PFOA Story among Those Highlighted at EPA Summit." June 25, 2018. https://www.benningtonbanner.com/stories/bennington-expert-speaks-at-pfoa-summit-in-new-hampshire,543172.

Time. 1954. "Business: Flight to the Suburbs." March 22, 1954. https://time.com/archive/6621959/business-flight-to-the-suburbs/

Time. 1970a. "Environment: Ecology of a Ghetto." April 6, 1970. https://time.com/archive/6877061/environment-ecology-of-a-ghetto/

Time. 1970b. "Environment: The Rise of Anti-Ecology." August 3, 1970. https://time.com/archive/6814404/environment-the-rise-of-anti-ecology/

Time. 1977. "The American Underclass." August 29, 1977. https://content.time.com/time/subscriber/article/0,33009,915331,00.html.

Toosi, Mitra. 2002. "A Century of Change: The U.S. Labor Force, 1950–2050." *Monthly Labor Review*, May 2002, 15–28. Bureau of Labor Statistics. https://www.bls.gov/opub/mlr/2002/05/art2full.pdf.

Towey, Hannah. 2022. "JPMorgan CEO Jamie Dimond: 'Why Can't We Get It Through Our Thick Skulls?' America Boosting Oil and Gas Production Is 'Not Against' Climate Change." *Business Insider*, August 14, 2022. https://www.businessinsider.com/jpmorgan-jamie-dimon-oil-gas-production-not-against-climate-change-2022-8.

Trump, Rosemary. 1980. Remarks at Safe Energy and Full Employment Conference. Environmentalists for Full Employment Records, 1969–1984, AIS.1984.27, Archives & Special Collections, University of Pittsburgh, Box 2, Folder 32.

Tuholske, Lilly. 1993. "Slick Marketing Ploys Helped Kill BTU Tax." *The Missoulian*, June 12, 1993.

Tupper, Margot. 1966. *No Place to Play*. Philadelphia: Chilton Books. .

United Auto Workers. 2023. "UAW President Fain and Vice President Browning on UAW Ford Tentative Agreement." October 29, 2023. YouTube video, 18:30–19:30. https://www.youtube.com/watch?v=cT9XpLEmy9U.

US Bureau of Labor Statistics. 2023. "Productivity Change in the Nonfarm Business Sector, 1947 Q1–2023 Q1." https://www.bls.gov/productivity/#:~:text=In%20manufacturing%2C%20productivity%20increased%200.2,labor%20costs%20increased%205.7%20percent.&text=Retail%20trade%20productivity%20increased%207.7,growth%20in%20nondurable%20goods%20industries.

US Census Bureau. 2020. "Current Population Survey, 1960 to 2020 Annual Social and Economic Supplement." https://www.census.gov/library/stories/2020/09/poverty-rates-for-blacks-and-hispanics-reached-historic-lows-in-2019.html.

US Department of Labor. 1965. "The Negro Family: The Case for National Action." Office of Policy Planning and Research. Washington, DC: Government Printing Office.

US Department of Labor. 2018. "Earnings and Ratio's." Women's Bureau. https://www.dol.gov/agencies/wb/data/facts-over-time/earnings-and-earnings-ratios.

US Energy Information Administration. 2020. "Today in Energy: US Annual Domestic Production and Foreign Imports of Uranium, 1950–2019." July 17, 2020. https://www.eia.gov/todayinenergy/detail.php?id=44416.

US Energy Information Administration. 2021. "U.S. Field Production of Crude Oil." https://www.eia.gov/dnav/pet/hist/LeafHandler.ashx?n=pet&s=mcrfpus2&f=a.

US Environmental Protection Agency. 1983. Economic Dislocation Early Warning System, 1980–1982 Quarterly Report. Washington, DC: National Service Center for Environmental Publications.

US Senate. 2019. "Manchin Presses DOE Officials for Update on Appalachian Storage Hub." Committee on Energy and Natural Resources. July 9, 2019. https://www.energy.senate.gov/2019/7/manchin-presses-doe-officials-for-update-on-appalachian-storage-hub.

US Tariff Commission. 1952. "Synthetic Organic Chemicals: US Production and Sales, 1951." Report No. 175. Washington DC: Government Printing Office.

US Tariff Commission. 1962. "Synthetic Organic Chemicals: US Production and Sales, 1961." TC Publication 72. Washington, DC: Government Printing Office.

Vance, J. D. 2016. "How Donald Trump Seduced America's White Working Class." *The Guardian*, September 10, 2016. https://www.theguardian.com/commentisfree/2016/sep/10/jd-vance-hillbilly-elegy-donald-trump-us-white-poor-working-class.

VanderHart, Dirk. 2020. "As Cap-and-Trade Bill Moves Forward, Republicans Head for Exits." Oregon Public Broadcasting." February 24, 2020. https://www.opb.org/news/article/oregon-republican-senator-walkout-cap-and-trade-bill/.

Vandewater, Bob. 1993. "Btu Tax Called Costly for State Business." *The Oklahoman*, April 24, 1993. https://www.oklahoman.com/story/news/1993/04/24/btu-tax-called-costly-for-state-business-energy-leaders-foresee-459-million-price-tag/62461287007/.

Various Authors. 2019. "Against the Dead Consensus." *First Things*, March 21, 2019. https://www.firstthings.com/web-exclusives/2019/03/against-the-dead-consensus.

Vice News. 2020. *Why Coal Country Elected Trump*. YouTube video, February 28, 2020. https://www.youtube.com/watch?v=EJZmr7E2M_U.

Viguerie, Richard. 1977. "Let's Get Union Members to Support Conservatives." *Conservative Digest* 3 (August).

Von Drehle, David. 2018. "Folks in the Midwest Have Trump All Figured Out." *Washington Post*, April 6, 2018. https://www.washingtonpost.com/opinions/folks-in-the-midwest-have-trump-all-figured-out/2018/04/06/52c7e9ce-39b9-11e8-8fd2-49fe3c675a89_story.html?noredirect=on&utm_term=.bc1a040e1708.

Waldman, Paul. 2016. "When Will White Working-Class Trump Voters See the Scam?" *Chicago Tribune*, November 25, 2016. https://www.chicagotribune.com/opinion/commentary/ct-trump-white-working-class-scam-20161125-story.html.

Washington Post. 2016. "Trump Receives Warm Welcome in Coal Country." May 6, 2016. https://www.washingtonpost.com/video/politics/trump-receives-warm-welcome-in-coal-country/2016/05/06/9259c5ea-1327-11e6-a9b5-bf703a5a7191_video.html.

Watt, James G. 1977. Notes in Preparation for Interview with the Mountain States Legal Foundation. Box 7, Folder 4. James G. Watt Papers, Collection Number 7667, American Heritage Center, University of Wyoming.

Watt, James G. 1982. "Ours Is the Earth." *Saturday Evening Post*, January/February. Box 6, Folder 12. James G. Watt Papers, Collection Number 7667, American Heritage Center, University of Wyoming.

Weidenbaum, Murray. 1980. "Public Policy: No Longer a Spectator Sport for Business." Washington University St Louis, Center for the Study of American Business, Publication Number 34.

White, Dustin. 2021. Interview with author. July 27, 2021.

White House. 2021. "Executive Order on Protecting Public Health and the Environment and Restoring Science to Tackle the Climate Crisis." January 20, 2021. https://www.whitehouse.gov/briefing-room/presidential-actions/2021/01/20/executive-order-protecting-public-health-and-environment-and-restoring-science-to-tackle-climate-crisis/.

Whitney, Peyton. 2023. "Number of Renters Burdened by Housing Costs Reached a Record High in 2021." Joint Center for Housing Studies of Harvard University. https://www.jchs.harvard.edu/blog/number-renters-burdened-housing-costs-reached-record-high-2021.

Winpisinger, William. 1980. "Remarks at Safe Energy and Full Employment Conference." Environmentalists for Full Employment Records, 1969–1984, AIS.1984.27, Archives & Special Collections, University of Pittsburgh, Box 2, Folder 32.

WorldData. 2023. "Inflation Rates in the United States of America." https://www.worlddata
.info/america/usa/inflation-rates.php#:~:text=Development%20of%20inflation%20rates%20
in,rate%20was%203.8%25%20per%20year.

World Meteorological Organization. 2024. "Climate Change Indicators Reached Record Levels in
2023." March 19, 2024. https://wmo.int/news/media-centre/climate-change-indicators-reached
-record-levels-2023-wmo.

Zimmerman, Sarah, and Gillian Flaccus. 2019. "Oregon GOP Senator on Governor Sending
Police: 'Send Bachelors and Come Heavily Armed.'" KVAL, June 20, 2019. https://kval.com/news
/local/oregon-senator-on-governor-sending-police-send-bachelors-and-come-heavily-armed.

SECONDARY SOURCES

Allen, Theodore W. 2001. "On Roediger's 'Wages of Whiteness.'" *Cultural Logic: A Journal of
Marxist Theory & Practice* 8. doi:https://doi.org/10.14288/clogic.v8i0.191856.

Allen, Theodore W. 2021. *The Invention of the White Race: The Origin of Racial Oppression.* London:
Verso.

Anderson, Carol. 2017. *White Rage.* New York and London: Bloomsbury.

Andrews, David, and Olga Naidenko. 2020. "Population-Wide Exposure to Per- and Polyflou-
roalkyl Substances from Drinking Water." *Environmental Science & Technology Letters* 7 (12):
931–936.

Arendt, Hannah. 1973 [1951]. *The Origins of Totalitarianism.* Vol. 244. Boston: Houghton Mifflin Harcourt.

Arendt, Hannah. 1998 [1958]. *The Human Condition.* Chicago: University of Chicago Press.

Arnesen, Eric. 2001. "Whiteness and the Historians' Imagination." *International Labor and
Working-Class History* 60:3–32.

Aronoff, Kate, Alyssa Battistoni, Daniel Cohen and Thea Riofrancos. 2019. *A Planet to Win: Why
We Need a Green New Deal.* London: Verso.

Aronoff, Kate. 2021. *Over-Heated: How Capitalism Broke the Planet and How We Fight Back.* New
York: Bold Types Books.

Aronoff, Kate. 2022a. "The Conservative Plot against Green Investment." *New Republic*, January 4.
https://newrepublic.com/article/164916/alec-esg-fossil-fuel-investment

Aronoff, Kate. 2022b. "Elon Musk Is the Newest Acolyte of the Right's Critical Energy Theory
Nonsense." *New Republic*, May 23, 2022. https://newrepublic.com/article/166561/elon-musk
-newest-acolyte-rights-critical-energy-theory-nonsense.

Arrighi, Giovanni. 1994. *The Long Twentieth Century: Money, Power, and the Origins of Our Times.*
London: Verso.

Baer, Kenneth. 2000. *Reinventing Democrats: The Politics of Liberalism from Reagan to Clinton.*
Lawrence: University Press of Kansas.

Bagley, William Chandler. 1942. "Soil Exhaustion and the Civil War." Washington DC: American Council on Public Affairs.

Baker, Andrew. 2018. *Bulldozer Revolutions: A Rural History of the Metropolitan South*. Athens, GA: University of Georgia Press.

Bakker, Isabella. 2007. "Social Reproduction and the Constitution of a Gendered Political Economy." *New Political Economy* 12 (4): 541–556.

Barca, Stefania. 2012. "On Working-Class Environmentalism: A Historical and Transnational Overview." *Interface* 4 (2): 61–80.

Barca, Stefania. 2014. "Laboring the Earth: Transnational Reflections on the Environmental History of Work." *Environmental History* 19 (1): 3–27.

Bartley, Numan. 1997 [1969]. *The Rise of Massive Resistance: Race and Politics in the South during the 1950s*. Baton Rouge: Louisiana State University Press.

Bartels, Larry. 2020. "Ethnic Antagonism Erodes Republicans Commitments to Democracy." *Proceedings of the National Academy of Science* 117 (37): 22752–22759.

Battistoni, Alyssa. 2017a. "Living, Not Just Surviving." *Jacobin*, August 15, 2017. https://www.jacobinmag.com/2017/08/living-not-just-surviving/.

Battistoni, Alyssa. 2017b. "Bringing in the Work of Nature: From Natural Capital to Hybrid Labor." *Political Theory* 45 (1): 5–31.

Baumgartner, Frank, and Bryan Jones. 1991. "Agenda Dynamics and Policy Subsystems." *Journal of Politics* 53 (4): 1044–1074.

Bazzi, Samuel, Andreas Ferrara, Martin Fiszbein, Thomas Pearson and Patrick Testa. 2023. "The Other Great Migration: Southern Whites and the New Right." National Bureau of Economic Research Working Paper 29506. https://www.nber.org/papers/w29506.

Beck, Ulrich. 1992. *Risk Society: Towards a New Modernity*. Vol. 17. London: Sage.

Becker, Gary. 1993. *A Treatise on the Family*. Cambridge, MA: Harvard University Press.

Beckert, Sven. 2015. *Empire of Cotton: A Global History*. New York: Vintage.

Beltrán, Cristina. 2020. *Cruelty as Citizenship: How Migrant Suffering Sustains White Democracy*. Minneapolis: University of Minnesota Press.

Benegal, Salil. 2018. "The Spillover of Race and Racial Attitudes into Public Opinion about Climate Change." *Environmental Politics* 27 (4): 733–756.

Benjamin, Walter. 2003. *Selected Writings, 1938–1940*. Vol. 4. Edited by H. Eiland and M. W. Jennings. Cambridge, MA: Harvard University Press.

Berger, Bennett. 1960. *Working-Class Suburb: A Study of Auto Workers in Suburbia*. Berkeley: University of California Press.

Berry, Daina Ramey. 2017. *The Price for Their Pound of Flesh: The Value of the Enslaved, from Womb to Grave, in the Building of a Nation*. Boston: Beacon Press.

Bessire, Lucas, and David Bond. 2014. "Ontological Anthropology and the Deferral of Critique." *American Ethnologist* 41 (3): 440–456.

Bhardwaj, Ankit. 2023. "The Soils of Black Folk: WEB Du Bois's Theory of Environmental Racialization." *Sociological Theory* 41 (2): 105–128.

Bhattacharya, Tithi, ed. 2017. *Social Reproduction Theory: Remapping Class, Recentering Oppression.* London: Pluto Press.

Black, Megan. 2018. *The Global Interior: Mineral Frontiers and American Power.* Cambridge, MA: Harvard University Press.

Blackhawk, Ned. 2006. *Violence over the Land: Indians and Empires in the Early American West.* Cambridge, MA: Harvard University Press.

Block, Fred. 1987. *Revising State Theory: Essays in Politics and Postindustrialism.* Philadelphia: Temple University Press.

Bloom, Jack. 1987. *Class, Race, & The Civil Rights Movement.* Bloomington: Indiana University Press.

Bluestone, Barry, and Bennett Harrison. 1982. *The Deindustrialization of America.* New York: Basic Books.

Blum, Elizabeth. 2002. "Power, Danger, and Control: Slave Women's Perceptions of Wilderness in the Nineteenth Century." *Women's Studies* 31 (2): 247–265.

Blumberg, Leonard, and Michael Lalli. 1966. "Little Ghettoes: A Study of Negroes in the Suburbs." *Phylon* 27 (2): 117–131.

Boggs, James. 1963. *The American Revolution: Pages from a Negro Worker's Notebook.* https://libcom.org/article/american-revolution-pages-negro-workers-notebook.

Bond, David. 2021. "Contamination in Theory and Protest." *American Ethnologist* 48 (4): 386–403.

Bond, David. 2022. *Negative Ecologies: Fossil Fuels and the Discovery of the Environment.* Oakland: University of California Press.

Bor, Jacob. 2017. "Diverging Life Expectancies and Voting Patterns in the 2016 US Presidential Election." *American Journal of Public Health* 107 (10): 1560–1562. https://ajph.aphapublications.org/doi/10.2105/AJPH.2017.303945

Bosworth, Kai. 2022. *Pipeline Populism: Grassroots Environmentalism in the Twenty-First Century.* Minneapolis: University of Minnesota Press.

Brick, Phil. 1995. "Determined Opposition: The Wise Use Movement Challenges Environmentalism." *Environment* 37 (8): 17–42.

Bridges, Amy. 1986. "Becoming American: The Working-Classes in the United States before the Civil War." In *Working-Class Formation: Nineteenth Century Patterns in Western Europe and the United States*, edited by Katznelson and Zolberg, 157–196. Princeton, NJ: Princeton University Press.

Brodkin, Karen. 1998. *How Jews Became White Folks and What That Says about Race in America.* New Brunswick, NJ: Rutgers University Press.

Brown, Alleen. 2018. "The Infiltrator: How an Undercover Oil Industry Mercenary Tricked Pipeline Opponents into Believing He Was One of Them." *The Intercept*, December 30, 2018. https://theintercept.com/2018/12/30/tigerswan-infiltrator-dakota-access-pipeline-standing-rock/.

Brown, Wendy. 2015. *Undoing the Demos: Neoliberalism's Stealth Revolution*. New York: Zone Books.

Brown, Wendy. 2018. "Who Is Not a Neoliberal Today." *Tocqueville21 Democracy and Politics* (blog), January 18, 2018. https://tocqueville21.com/interviews/wendy-brown-not-neoliberal-today/

Brown, Wendy. 2019. *In the Ruins of Neoliberalism: The Rise of Anti-Democratic Politics in the West*. New York: Columbia University Press.

Brulle, Robert J. 2014. "Institutionalizing Delay: Foundation Funding and the Creation of US Climate Change Counter-Movement Organizations." *Climatic Change* 122 (4): 681–694.

Brulle, Robert J. 2023. "Advocating Inaction: A Historical Analysis of the Global Climate Coalition." *Environmental Politics* 32 (2): 185–206.

Bruyneel, Kevin. 2021. *Settler Memory: The Disavowal of Indigeneity and the Politics of Race in the United States*. Chapel Hill: University of North Carolina Press.

Bullard, Robert. 1990. *Dumping in Dixie: Race, Class, and Environmental Quality*. New York: Routledge.

Bullard, Robert, ed. 1993. *Confronting Environmental Racism: Voices from the Grassroots*. Boston: South End Press.

Burawoy, Michael. 1985. *The Politics of Production: Factory Regimes under Capitalism and Socialism*. London: Verso.

Burke, James, Gerald Epstein, and Minsik Choi. 2004. "Rising Foreign Outsourcing and Employment Losses in U.S. Manufacturing, 1987–2002." Political Economy Research Institute, University of Massachusetts Amherst.

Burley, Shane. 2017. *Fascism Today: What It Is and How to End It*. Chico, CA: AK Press.

Carnes, Nicholas, and Noam Lupu. 2017. "It's Time to Bust the Myth: Most Trump Voters Were Not Working-Class." *Washington Post*, Monkey Cage. June 5, 2017. https://www.washingtonpost.com/news/monkey-cage/wp/2017/06/05/its-time-to-bust-the-myth-most-trump-voters-were-not-working-class/?utm_term=.5a9732daad8b.

Case, Ann, and Angus Deaton. 2021. Deaths of Despair and the Future of Capitalism. New Brunswick, NJ: Princeton University Press.

Catte, Elizabeth. 2019. "Why Trump Country Isn't as Republican as You Think." *The Guardian*, February 22, 2019. https://www.theguardian.com/news/2019/feb/22/trump-country-republican-appalachia-virginia-activism.

Cawley, R. McGreggor. 1993. *Federal Land, Western Anger*. Lawrence: University Press of Kansas.

Center for Working-Class Politics. 2021. "Commonsense Solidarity: How a Working-Class Coalition Can Be Built, and Maintained." https://www.workingclasspolitics.org/press-publications.

Cha, J. Mijin, Dimitris Stevis, Todd Vachon, Vivian Price, and Maria Brescia-Weiler. 2022. "A Green New Deal for All: The Centrality of a Worker and Community Led Just Transition in the US." *Political Geography* 95:102594.

Chinni, Dante. 2017. "Trump County Voters Aren't Downtrodden, but They Are Being Left Behind." *NBC News*, July 24, 2017. https://www.nbcnews.com/politics/donald-trump/trump -county-voters-left-behind-not-downtrodden-n786056.

Coates, Ta-Nehisi. 2017. "The First White President." *The Atlantic* 320 (3): 74–87.

Cohen, Lizabeth. 2004. *A Consumers' Republic: The Politics of Mass Consumption in Postwar America*. New York: Vintage Books.

Cohen, William. 1991. *At Freedom's Edge: Black Mobility and the Southern White Quest for Racial Control, 1861–1915*. Baton Rouge: Louisiana State University Press.

Coles, Romand. 2016. *Visionary Pragmatism: Radical and Ecological Democracy in Neoliberal Times*. Durham, NC: Duke University Press.

Cowie, Jefferson. 2010. *Stayin' Alive: The 1970s and the Last Days of the Working-Class*. New York: The New Press.

Cox, Daniel, and Robert Jones. 2016. "Still Live near Your Hometown? If You're White, You're More Likely to Support Trump." PRII/The Atlantic Survey, October 6, 2016. https://www.prri .org/research/prri-atlantic-oct-6-poll-politics-election-clinton-trump/.

Cox, Oliver Cromwell. 1945. "An American Dilemma: A Mystical Approach to the Study of Race Relations." *Journal of Negro Education* 14 (2): 132–148.

Cramer, Katherine. 2016. *The Politics of Resentment: Rural Consciousness in Wisconsin and the Rise of Scott Walker*. Chicago: University of Chicago Press

Cronon, William. 1983. *Changes in the Land: Indians, Colonists, and the Ecology of New England*. New York: Hill and Wang.

Cronon, William. 1992. *Nature's Metropolis: Chicago and the Great West*. New York: W.W. Norton & Company.

Daggett, Cara. 2018. "Petro-Masculinity: Fossil Fuels and Authoritarian Desire." *Millennium: Journal of International Studies* 47 (1): 25–44.

Dalla Costa, Mariarosa, and Selma James. 1972. *The Power of Women and the Subversion of Community*. Bristol, UK: Falling Wall Press.

Davis, Mike. 1986. *Prisoners of the American Dream*. London: Verso.

Dewey, Scott. 1998. "Working for the Environment: Organized Labor and the Origins of Environmentalism in the United States, 1948–1970." *Environmental History* 3 (1): 45–63.

DiAngelo, Robin. 2018. *White Fragility: Why It's So Hard for White People to Talk about Racism*. Boston: Beacon Press.

Diouf, Sylviane. 2014. *Slavery's Exiles: The Story of the American Maroons*. New York: New York University Press.

Dixon, Melvin. 1987. *Ride Out the Wilderness: Geography and Identity in Afro-American Literature*. Urbana: University of Illinois Press.

Dray, Philip. 2011. *There Is Power in a Union: The Epic Story of Labor in America*. New York: Anchor Books.

Dreyfuss, Robert. 2001. "How the DLC Does It." *American Prospect*, April 23, 2001

Du Bois, W. E. B. 1992 [1935]. *Black Reconstruction in America, 1860–1880*. New York: The Free Press.

Dunbar-Ortiz, Roxanne. 2014. *An Indigenous Peoples' History of the United States*. Vol. 3. Boston: Beacon Press.

Dunlap, Riley, and Aaron McCright. 2010. "Climate Change Denial: Sources, Actors, and Strategies." In *Routledge Handbook of Climate Change and Society*, edited by Constance Lever-Tracy, 240–259. London: Routledge.

Dwyer, Rachel. 2013. "The Care Economy? Gender, Economic Restructuring, and Job Polarization in the U.S. Labor Market." *American Sociological Review* 78 (3): 390–416.

Ehrenreich, Barbara. 1989. *Fear of Falling: The Inner Life of the Middle Class*. New York: HarperPerennial.

Enders, Adam, and Jamil Scott. 2019. "The Increasing Racialization of American Electoral Politics, 1988–2016." *American Politics Research* 47 (2): 275–303.

Erlandson, Dawn. 1994. "The BTU Tax Experience: What Happened and Why It Happened." *Pace Environmental Law Review* 12:173.

Esposito, John. 1970. *Vanishing Air: Ralph Nader's Study Group Report on Air Pollution*. New York: Grossman Publishers.

Estes, Nick. 2019. *Our History Is the Future*. London: Verso.

Farley, Reynolds. 1970. "The Changing Distribution of Negroes within Metropolitan Areas: The Emergence of Black Suburbs." *American Journal of Sociology* 75 (4): 512–529.

Farrell, Justin. 2016. "Corporate Funding and Ideological Polarization about Climate Change." *Proceedings of the National Academy of Sciences* 113 (1): 92–97.

Farrell, Justin. 2021. *Billionaire Wilderness: The Ultra-Wealthy and the Remaking of the American West*. Princeton, NJ: Princeton University Press.

Federici, Silvia. 1974. *Wages Against Housework*. Bristol, UK: Power of Women Collective and Falling Wall Press.

Federici, Silvia. 2012. *Revolution at Point Zero*. Oakland: PM Press.

Federici, Silvia. 2019. "Social Reproduction Theory: History, Issues and Present Challenges." *Radical Philosophy* 2 (3): 55–57.

Fiege, Mark. 2012. *The Republic of Nature: An Environmental History of the United States*. Seattle: University of Washington Press.

Fields, Barbara. 2001. "Whiteness, Racism, and Identity." *International Labor and Working-Class History* 60: 48–56.

Finney, Carolyn. 2014. *Black Faces, White Spaces: Reimagining the Relationship of African Americans to the Great Outdoors.* Chapel Hill: University of North Carolina Press.

Fishman, Robert. 1987. *Bourgeois Utopias: The Rise and Fall of Suburbia.* New York: Basic Books.

Foley, Neil. 1997. *The White Scourge: Mexicans, Blacks, and Poor Whites in Texas Cotton Culture.* Berkeley: University of California Press.

Foner, Eric. 1995 [1970]. *Free Soil, Free Labor, Free Men.* Oxford: Oxford University Press.

Foner, Eric. 2007. *Nothing but Freedom: Emancipation and Its Legacy.* Baton Rouge. LA: LSU Press.

Foster, John Bellamy. 2002 [1993]. "The Limits of Environmentalism without Class: Lessons from the Ancient Forest Struggle in the Pacific Northwest." In *Ecology against Capitalism.* London: Monthly Review Books.

Foster, John Bellamy, Brett Clark, and Richard York. 2010. *The Ecological Rift: Capitalism's War on the Earth.* New York: Monthly Review Press.

Foucault, Michel. 2008. *The Birth of Biopolitics: Lectures at the College de France, 1978–79.* New York: Palgrave MacMillan.

Fraser, Nancy. 2014. "Can Society Be Commodities All the Way Down? Post-Polanyian Reflections on Capitalist Crisis." *Economy and Society* 43 (4): 541–558.

Fraser, Nancy. 2017. "Crisis of Care: On the Social Reproductive Contradictions of Contemporary Capitalism." In *Social Reproduction Theory: Remapping Class, Recentering Oppression,* edited by Tithi Bhattacharya, 21–36. London: Pluto Press.

Fraser, Steve. 2016. *The Limousine Liberal: How an Incendiary Image United the Right and Fractured America.* New York: Basic Books.

Frieden, Jeffrey. 2006. *Global Capitalism: Its Fall and Rise in the Twentieth Century.* New York: W.W. Norton.

Fromm, Erich. 2002 [1955]. *The Sane Society.* New York: Routledge.

Galbraith, John Kenneth. 1998 [1958]. *The Affluent Society.* New York: Houghton Mifflin.

Gates, Robbins. 1964. *The Making of Massive Resistance: Virginia's Politics of Public School Desegregation, 1954–1956.* Chapel Hill: University of North Carolina Press.

Gedicks, Al. 2001. *Resource Rebels: Native Challenges to Mining and Oil Corporations.* Cambridge, MA: South End Press.

Gedicks, Al. 2018. "Wisconsin's 'Standing Rock': The Proposed Back Forty Mine." *Race and Class* 60 (2): 106–113.

Geismer, Lily. 2015. *Don't Blame Us: Suburban Liberals and the Transformation of the Democratic Party.* Princeton, NJ: Princeton University Press.

Germic, Stephen. 2001. *American Green: Class, Crisis, and the Deployment of Nature in Central Park, Yosemite and Yellowstone*. Lanham, MD: Lexington Books.

Gest, Justin. 2016. *The New Minority: White Working-Class Politics in an Age of Immigration and Inequality*. Oxford: Oxford University Press.

Glave, Dianne, and Mark Stoll. 2006. *To Love the Wind and the Rain: African Americans and Environmental History*. Pittsburgh: University of Pittsburgh Press.

Golden, Kathryn Benjamin. 2021. "'Armed in the Great Swamp': Fear, Maroon Insurrection, and the Insurgent Ecology of the Great Dismal Swamp." *Journal of African American History* 106 (1): 1–26.

Goldin, Claudia, and Robert A. Margo. 1992. "The Great Compression: The Wage Structure in the United States at Mid-Century." *Quarterly Journal of Economics* 107 (February 1992): 1–34.

Gómez, Laura. 2007. *Manifest Destinies: The Making of the Mexican American Race*. New York: NYU Press.

Goodwin, James, Yong-Fang Kuo, David Brown, David Juurlink, and Mukaila Raji. 2016. "Association of Chronic Opioid Use with Presidential Voting Patterns in US Counties in 2016." *Journal of the American Medical Association Network Open*. https://jamanetwork.com/journals/jamanetworkopen/fullarticle/2685627.

Goodwin, Jeff. 2022. "Black Reconstruction as Class War." *Catalyst: A Journal of Theory & Strategy* 6 (1): 52–95.

Gordon, Robert. 1998. "'Shell No!' OCAW and the Labor-Environmental Alliance." *Environmental History* 3 (4): 460–487.

Gordon, Robert. 2004. "Environmental Blues: Working-Class Environmentalism and the Labor-Environmental Alliance, 1968–1985." PhD diss., Wayne State University.

Gorz, Andre. 1982 [1980]. *Farewell to the Working-Class: An Essay on Post-Industrial Socialism*. Boston: South End Press.

Gottlieb, Robert. 2005. *Forcing the Spring: The Transformation of the American Environmental Movement*. Washington, DC: Island Press.

Gould, Kenneth, Allan Schnaiberg, and Adam Weinberg. 1996. *Local Environmental Struggles: Citizen Activism in the Treadmill of Production*. Cambridge, UK: Cambridge University Press.

Gould, Kenneth, David Pellow, and Allan Schnaiberg. 2004. "Interrogating the Treadmill of Production: Everything You Wanted to Know About the Treadmill but Were Afraid to Ask." *Organization & Environment* 17 (3): 296–316.

Gould, Kenneth, David Pellow, and Allan Schnaiberg. 2008. *The Treadmill of Production*. Boulder: Paradigm Publishers.

Gouldner, Alvin. 1979. *The Future of Intellectuals and the Rise of the New Class*. New York: Seabury.

Graf, William. 1990. *Wilderness Preservation and the Sagebrush Rebellions*. Savage, MD: Rowman & Littlefield.

Grandin, Greg. 2019. *The End of the Myth: From the Frontier to the Border Wall in the Mind of America*. New York: Metropolitan Books.

Grasso, Marco. 2019 "Oily Politics: A Critical Assessment of the Oil and Gas Industry's Contribution to Climate Change." *Energy Research & Social Science* 50:106–115.

Gregory, James. 2005. *The Southern Diaspora: How the Great Migrations of Black and White Southerners Transformed America*. Chapel Hill, NC: University of North Carolina Press.

Grossman, Zoltan. 2017. *Unlikely Alliances: Native Nations and White Communities Join to Defend Rural Lands*. Seattle: University of Washington Press.

Grossman, Zoltan. 2019. "Populist Alliances of 'Cowboys and Indians' Are Protecting Rural Lands." *The Conversation,* May 16. https://theconversation.com/populist-alliances-of-cowboys-and-indians-are-protecting-rural-lands-114268.

Grundman, Adolph. 1972. "Public School Desegregation in Virginia from 1954 to the Present." PhD diss., Wayne State University. Paper 952.

Guastella, Dustin. 2022. "The Left Is Still Losing the Working-Class." *Jacobin,* September 7. https://jacobin.com/2022/09/the-left-is-still-losing-the-working-class.

Habermas, Jurgen. 1989 [1962]. *The Structural Transformation of the Public Sphere: An Inquiry into a Category of Bourgeois Society*. Cambridge, MA: MIT Press.

Hahn, Steven. 1982. "Hunting, Fishing, and Foraging: Common Rights and Class Relations in the Postbellum South." *Radical History Review* 26: 37–64

Haider, Asad. 2018. *Mistaken Identity: Race and Class in the Age of Trump*. London: Verso.

Hale, Grace Elizabeth. 1999. *Making Whiteness: The Culture of Segregation in the South, 1890–1940*. New York: Vintage.

Hale, Jon. 1995. "The Making of the New Democrats." *Political Science Quarterly* 110 (2): 207–232.

Hare, Nathan. 1971. "Black Ecology." *Trends* 8 (3): 4–8.

Harris, Cheryl. 1993. "Whiteness as Property." *Harvard Law Review* 106 (8): 1707–1791.

Hartman, Saidiya. 2016. "The Belly of the World: A Note on Black Women's Labors." *Souls* 18 (1): 166–173.

Hartmann, Heidi. 1979. "The Unhappy Marriage of Marxism and Feminism: Towards a More Progressive Union." *Capital and Class* 3 (2): 1–33.

Harvard Law Review. 2014. "Citizens United at Work: How the Landmark Decision Legalized Political Coercion in the Workplace." Notes. *Harvard Law Review* 128: 669–690.

Hayek, Friedrich. 1994 [1944]. *The Road to Serfdom*. Chicago: University of Chicago Press.

Hays, Samuel. 1987. *Beauty, Health, and Permanence: Environmental Politics in the United States, 1955–85*. Cambridge, UK: Cambridge University Press.

Helvarg, David. 1994. *War against the Greens*. San Francisco: Sierra Club Books.

Hertel-Fernandez, Alexander. 2020. "What Americans Think about Worker Power and Organization: Lessons from a New Survey." Data for Progress. https://www.filesforprogress.org/memos/worker-power.pdf.

Heynen, Nik. 2018. "Toward an Abolition Ecology." *Abolition: A Journal of Insurgent Politics* 1: 240–247.

Higginson, Thomas Wentworth. 1870. *Army Life in a Black Regiment.* Boston: Fields, Osgood & Company.

Hochschild, Arlie. 2016. *Strangers in Their Own Land: Anger and Mourning on the American Right.* New York: The New Press.

Hornborg, Alf. 1998. "Ecosystems and World Systems: Accumulation as an Ecological Process." *Journal of World-Systems Research* 4: 169–177.

Hornborg, Alf. 2006. "Footprints in the Cotton Fields: The Industrial Revolution as Time–Space Appropriation and Environmental Load Displacement." *Ecological Economics* 59 (1): 74–81.

Hosang, Daniel Martinez, and Joseph Lowndes. 2019. *Producers, Parasites, Patriots: Race and the New Right-Wing Politics of Precarity.* Minneapolis: University of Minnesota Press.

Hough, Franklin. 1873. "On the Duty of Governments in the Preservation of Forests." *Proceedings of the American Association for the Advancement of Science.* Library of Congress. https://www.loc.gov/resource/gdclccn.12018878/?sp=1.

Huber, Matthew. 2013. *Lifeblood: Oil, Freedom and the Forces of Capital.* Minneapolis: University of Minnesota Press.

Huber, Matthew. 2019. "Ecological Politics for the Working-Class." *Catalyst* 3 (1): 7–45.

Huber, Matthew. 2022. *Climate Change as Class War: Building Socialism on a Warming Planet.* London: Verso.

Hughes, Donald J. 1985. "Theophrastus as Ecologist." *Environmental Review* 9 (4): 296–306.

Hultgren, John. 2015. *Border Walls Gone Green: Nature and Anti-Immigrant Politics in America.* Minneapolis: University of Minnesota Press.

Hultgren, John, and Dimitris Stevis. 2020. "Interrogating Socio-Ecological Coalitions: Environmentalist Engagements with Labor and Immigrants' Rights in the United States." *Environmental Politics* 29 (3): 457–478.

Hurley, Andrew. 1995. *Environmental Inequalities: Class, Race, and Industrial Pollution in Gary, Indiana, 1945–1980.* Chapel Hill: University of North Carolina Press.

Ignatiev, Noel. 1995. *How the Irish Became White.* New York: Routledge.

Ignatin, Noel, and Theodore Allen. 1969 [1967]. "White Blindspot." NYC Revolutionary Youth Movement. https://www.marxists.org/history/erol/ncm-1/whiteblindspot.pdf.

Inglehart, Ronald, and Jacques-Rene Rabier. 1986. "Political Realignment in Advanced Industrial Society: From Class-Based Politics to Quality of Life Politics." *Government and Opposition* 21 (4): 456–479.

Jackson, Kenneth. 1985. *Crabgrass Frontier: The Suburbanization of the United States*. New York: Oxford University Press.

Jacobin Editors. 2021. "Commonsense Solidarity: How a Working-Class Coalition Can Be Built, and Maintained." *Jacobin*, November 9. https://jacobin.com/2021/11/common-sense -solidarity-working-class-voting-report#:~:text=Strategy-,Commonsense%20Solidarity%3A%20 How%20a%20Working%2DClass%20Coalition,Can%20Be%20Built%2C%20 and%20Maintained&text=An%20experimental%20study%2C%20the%20first,on%20 working%2Dclass%20political%20views.

Jacoby, Karl. 2001. *Crimes Against Nature: Squatters, Poachers, Thieves and the Hidden History of American Conservation*. Berkeley: University of California Press.

Jardina, Ashley, and Robert Mickey. 2022. "White Racial Solidarity and Opposition to American Democracy." *Annals of the American Academy of Political and Social Science* 699 (1): 79–89.

Jacques, Peter J. 2012. "A General Theory of Climate Denial." *Global Environmental Politics* 12 (2): 9–17.

Jacques, Peter J., Riley E. Dunlap, and Mark Freeman. 2008. "The Organisation of Denial: Conservative Think Tanks and Environmental Scepticism. *Environmental Politics* 17 (3): 349–385.

Jerolmack, Colin. 2021. "This Could be the Start of a Rural Anti-Fracking Coalition." *New Republic*, May 17, 2021. https://newrepublic.com/article/162408/rural-anti-fracking-coalition

Jerolmack, Colin, and Edward Walker. 2018. "Please in My Backyard: Quiet Mobilization in Support of Fracking in an Appalachian Community." *American Journal of Sociology* 124 (2): 479–516.

Jessop, Bob. 2002. *The Future of the Capitalist State*. London: Polity.

Jobson, Ryan. 2021. "Dead Labor: On Racial Capital and Fossil Capital." In *Histories of Racial Capitalism*, edited by Destin Jenkins and Justin Leroy, 215–230. New York: Columbia University Press.

Johnson, Cedric. 2017. "The Panthers Can't Save Us Now." *Catalyst* 1 (1): 56–85.

Johnson, Cedric. 2019. "The Wages of Roediger: Why Three Decades of Whiteness Studies Has Not Produced the Left We Need." Nonsite.org 29. https://nonsite.org/the-wages-of-roediger-why -three-decades-of-whiteness-studies-has-not-produced-the-left-we-need/

Johnson, Cedric. 2023. *After Black Lives Matter: Policing and Anti-Capitalist Struggle*. London: Verso.

Johnson, Walter. 2013. *River of Dark Dreams*. Cambridge, MA: Harvard University Press.

Jowett, Benjamin. 1885. *The Politics of Aristotle*. Oxford, UK: Clarendon Press, xxii–xxv.

Karp, Matthew. 2023. "We Can't Ignore Class Dealignment." *Jacobin*, February 5, 2023. https:// jacobin.com/2023/02/matt-karp-class-dealignment-american-left-politics-blue-collar-voters

Katz, Jonathan. 2021. "It Happened Here." *Foreign Policy*, January 9, 2021. https://foreignpolicy .com/2021/01/09/capitol-riot-united-states-imperialism-trump/

Kazin, Michael. *The Populist Persuasion: An American History*. Ithaca: Cornell University Press.

Kazis, Richard, and Richard Grossman. 1982 [1991]. *Fear at Work: Job Blackmail, Labor and the Environment*. Philadelphia, PA: New Society Publishers.

Kelley, Robin D. G. 1999. "Building Bridges: The Challenge of Organized Labor in Communities of Color." *New Labor Forum* 5:42–58.

Ketcham, Christopher. 2015. "The Great Republican Land Heist." *Harpers* 330: 23–31.

Kimmerer, Robin Wall. 2013. *Braiding Sweetgrass: Indigenous Wisdom, Scientific Knowledge, and the Teachings of Plants*. Minneapolis: Milkweed Editions.

King, Desmond, and Rogers Smith. 2005. "Racial Orders in American Political Development." *American Political Science Review* 99 (1): 75–92.

Klein, Naomi. 2015. *This Changes Everything: Capitalism Versus the Climate*. New York: Simon and Schuster.

Kolko, Gabriel. 1955. "The CIO Faces Automation." *Dissent*. Fall 1955: 369–376.

Kosek, Jake. 2006. *Understories: The Political Life of Forests in Northern New Mexico*. Durham: Duke University Press.

Krippner, Greta. 2005. "The Financialization of the American Economy." *Socio-Economic Review* 3:173–208.

Kruse, Kevin. 2005. *White Flight: Atlanta and the Making of Modern Conservatism*. Princeton, NJ: Princeton University Press.

Kuttner, Robert. 2017. "The Man from Red Vienna." *New York Review of Books*, December 21, 2017. https://www.nybooks.com/articles/2017/12/21/karl-polanyi-man-from-red-vienna/

LaDuke, Winona. 1999. *All Our Relations: Native Struggles for Land and Life*. Cambridge, MA: South End Press.

Lassiter, Matthew. 2006. *The Silent Majority: Suburban Politics in the Sunbelt South*. Princeton, NJ: Princeton University Press.

Latour, Bruno. 1993. *We Have Never Been Modern*. Cambridge, MA: Harvard University Press.

Layzer, Judith. 2012. *Open for Business: Conservatives' Opposition to Environmental Regulation*. Cambridge, MA: MIT Press.

Layzer, Judith. 2016. *The Environmental Case*. 4th ed. Washington, DC: CQ Press.

Lee, Charles. 1993. "Beyond Toxic Wastes and Race." In *Confronting Environmental Racism: Voices from the Grassroots*, edited by Robert Bullard. Boston: South End Press.

Leopold, Les. 2007. *The Man Who Hated Work and Loved Labor: The Life and Times of Tony Mazzocchi*. White River Junction, VT: Chelsea Green Publishing.

Lester, James, David Allen, and Kelly Hill. 2001. *Environmental Injustice in the United States: Myths and Realities*. Boulder, CO: Westview Press.

Lichtenstein, Nelson. 1997. "Taft-Hartley: A Slave-Labor Law?" *Catholic University Law Review* 47 (3): 763–789.

Lichtenstein, Nelson. 2002. *State of the Union: A Century of American Labor.* Princeton, NJ: Princeton University Press.

Lichtenstein, Nelson. 2023. "UAW Strikers Have Scored a Historic, Transformative Victory." *Jacobin*, November 1. https://jacobin.com/2023/11/uaw-strike-contract-fain-victory.

Limerick, Patricia Nelson. 1988. *The Legacy of Conquest: The Unbroken Past of the American West.* New York: WW Norton & Company.

Lipsitz, George. 1998. *The Possessive Investment in Whiteness: How White People Profit from Identity Politics.* Philadelphia, PA: Temple University Press.

Litwack, Leon. 1961. *North of Slavery: The Negro in the Free States, 1790–1860.* Chicago: University of Chicago Press.

Loomis, Erik. 2016. *Empire of Timber: Labor Unions and the Pacific Northwest Forests.* Cambridge, UK: Cambridge University Press.

Lowndes, Joseph. 2008. *From the New Deal to the New Right: Race and the Southern Origins of Modern Conservatism.* New Haven: Yale University Press.

Lubell, Samuel. 1965. *The Future of American Politics.* New York: Harper Colophon Books.

Lupu, Noam, and Nocholas Carnes. 2021. "Trump Didn't Bring White Working-Class Voters to the Republican Party, He Kept Them Away." *Washington Post*, April 14. https://www.washingtonpost.com/politics/2021/04/14/trump-didnt-bring-white-working-class-voters-republican-party-data-suggest-he-kept-them-away/.

Lustgarten, Abrahm. 2020. "How Climate Migration Will Reshape America." *New York Times*, September 15. https://www.nytimes.com/interactive/2020/09/15/magazine/climate-crisis-migration-america.html.

Luxemburg, Rosa. 2003 [1913]. *The Accumulation of Capital.* New York: Routledge.

Malin, Stephanie. 2014. "There's No Real Choice but to Sign: Neoliberalisation and Normalization of Hydraulic Fracturing on Pennsylvania Farmland." *Journal of Environmental Studies and Sciences* 4 (1): 17–27.

Malm, Andreas. 2013. "The Origins of Fossil Capital: From Water to Steam in the British Cotton Industry." *Historical Materialism* 21 (1): 15–68.

Malm, Andreas. 2016. *Fossil Capital: The Rise of Steam Power and the Roots of Global Warming.* New York and London: Verso Books.

Malm, Andreas. 2018. "In Wildness is the Liberation of the World: On Maroon Ecology and Partisan Nature." *Historical Materialism* 26 (3): 3–37.

Malm, Andreas, and The Zetkin Collective. 2021. *White Skin, Black Fuel: On the Danger of Fossil Fascism.* New York: Verso.

Marable, Manning. 2007. *Race, Reform, and Rebellion: The Second Reconstruction and Beyond in Black America, 1945–2006.* Jackson: University Press of Mississippi.

Marsh, George Perkins. 2003 [1864]. *Man and Nature*. Seattle: University of Washington Press.

Marx, Karl. 1990 [1867]. *Capital, Volume 1*. London: Penguin Classics.

Marx, Karl, and Friedrich Engels. 1992 [1848]. *The Communist Manifesto*. Oxford and New York: Oxford University Press.

Massey, Douglass, and Nancy Denton. 1993. *American Apartheid: Segregation and the Making of the Underclass*. Cambridge, MA: Harvard University Press.

Mayer, Jane. 2017. *Dark Money: The Hidden History of the Billionaires behind the Rise of the Radical Right*. New York: Anchor Books.

McCright, Aaron M., and Riley E. Dunlap. 2011. "Cool Dudes: The Denial of Climate Change among Conservative White Males in the United States." *Global Environmental Change* 21 (4): 1163–1172.

McCright, Aaron M., and Riley E. Dunlap. 2013. "Bringing Ideology In: The Conservative White Male Effect on Worry about Environmental Problems in the USA." *Journal of Risk Research* 16 (2): 211–226.

McGirr, Lisa. 2001. *Suburban Warriors: The Origins of the New American Right*. Princeton, NJ: Princeton University Press.

McIntosh, Peggy. 1989. "White Privilege: Unpacking the Invisible Knapsack." *Peace and Freedom Magazine*, 10–12. Women's International League for Peace and Freedom.

McRae, Elizabeth Gillespie. 2018. *Mothers of Massive Resistance: White Women and the Politics of White Supremacy*. Oxford, UK: Oxford University Press.

Mellor, Mary. 1997. *Feminism & Ecology*. New York: New York University Press.

Mellor, Mary. 2006. "Ecofeminist Political Economy." *International Journal of Green Economics* 1 (1–2): 139–150.

Merchant, Carolyn. 1980. *The Death of Nature: Women, Ecology, and the Scientific Revolution*. San Francisco: Harper and Row.

Merchant, Carolyn. 1987. "The Theoretical Structure of Ecological Revolutions." *Environmental Review* 11 (4): 265–274.

Metzl, Jonathan. 2019. *Dying of Whiteness*. New York: Basic Books.

Meyer, John. 2015. *Engaging the Everyday: Environmental Social Criticism and the Resonance Dilemma*. Cambridge, MA: MIT Press.

Mies, Maria, and Vandana Shiva. 2014. *Ecofeminism*. London and New York: Zed Books.

Miliband, Ralph. 1969. *The State in Capitalist Society*. New York: Basic Books.

Mills, C. Wright. 2002 [1951]. *White Collar: The American Middle Classes*. Oxford, UK: Oxford University Press.

Mills, Charles W. 2004. "Racial Exploitation and the Wages of Whiteness." *The Changing Terrain of Race and Ethnicity*, edited by Maria Krysan and Amanda Lewis, 235–262. New York: Russell Sage Foundation.

Minchin, Timothy J. 2003. *Forging a Common Bond: Labor and Environmental Activism in the Base Lockout.* Gainesville: University Press of Florida.

Mitchell, Timothy. 2008. "Rethinking Economy." *Geoforum* 39 (3): 1116–1121.

Monbiot, George. 2016. "Neoliberalism—the Ideology at the Root of All Our Problems." *The Guardian*, April 15, https://www.theguardian.com/books/2016/apr/15/neoliberalism-ideology -problem-george-monbiot?CMP=share_btn_tw.

Montrie, Chad. 2008. *Making a Living: Work and Environment in the United States.* Chapel Hill: University of North Carolina Press.

Montrie, Chad. 2011. *A People's History of Environmentalism in the United States.* London and New York: Continuum.

Moody, Kim. 2000. "The Rank and File Strategy: Building a Socialist Movement in the U.S.." A Solidarity Working Paper. https://solidarity-us.org/rankandfilestrategy/

Moore, Jason. 2015. *Capitalism in the Web of Life: Ecology and the Accumulation of Capital.* New York: Verso.

Morena, Edouard, Dunja Krause, and Dimitris Stevis. 2020. *Just Transitions: Social Justice in a Low-Carbon World.* London: Pluto Press.

National Institute on Drug Abuse. 2023. "Drug Overdose Death Rates." National Institutes of Health, February 9. https://nida.nih.gov/research-topics/trends-statistics/overdose-death -rates.

Needham, Andrew. 2014. *Power Lines: Phoenix and the Making of the Modern Southwest.* Princeton, NJ: Princeton University Press.

Nelson, Bruce. 1996. "Class, Race and Democracy in the CIO: the New Labor History meets the 'Wages of Whiteness'." *International Review of Social History* 41 (3): 351–374.

Newman, William. 1957. "Americans in Subtopia." *Dissent.* Summer 1957. https://www .dissentmagazine.org/article/americans-in-subtopia/.

Nickerson, Michelle. 2012. *Mothers of Conservatism: Women and the Postwar Right.* Princeton: Princeton University Press.

Norgaard, Cari Marie. 2012. *Living in Denial: Climate Change, Emotions and Everyday Life.* Cambridge, MA: MIT Press.

Obach, Brian. 2004. *Labor and the Environmental Movement: The Quest for Common Ground.* Cambridge, MA: MIT Press.

Olmsted, Frederick Law. 1861. *The Cotton Kingdom: A Traveller's Observations on Cotton and Slavery in the American Slave States, 1853–1861.* New York: Mason Brothers.

Olson, Joel. 2004. *The Abolition of White Democracy.* Minneapolis: University of Minnesota Press.

Olson, Joel. 2008. "Whiteness and the Polarization of American Politics." *Political Research Quarterly* 61 (4): 704–718.

Omi, Michael, and Howard Winant. 1994. *Racial Formation in the United States*. New York: Routledge.

Oreskes, Naomi, and Erik Conway. 2010. *Merchants of Doubt*. New York: Bloomsbury Press.

Painter, Nell Irvin. 2010. *The History of White People*. New York: WW Norton & Company.

Painter, Nell Irvin. 2019. "What Is White America? The Identity Politics of the Majority." *Foreign Affairs* 98 (6).

Park, Lisa Sun-Hee, and David N. Pellow. 2004. "Racial Formation, Environmental Racism, and the Emergence of Silicon Valley." *Ethnicities* 4 (3): 403–424.

Patel, Raj, and Jason Moore. 2017. *A History of the World in Seven Cheap Things*. Oakland: University of California Press.

Paxton, Robert. 2021. "I've Hesitated to Call Donald Trump a Fascist. Until Now." *Newsweek*, January 11. https://www.newsweek.com/robert-paxton-trump-fascist-1560652

Peters, Margaret. 2008. *Conserving the Commonwealth: The Early Years of the Environmental Movement in Virginia*. Charlottesville: University of Virginia Press.

Phillips-Fein, Kim. 2009. *Invisible Hands: The Businessmen's Crusade Against the New Deal*. New York: W.W. Norton.

Piketty, Thomas. 2014. *Capital in the Twenty-First Century*. Cambridge, MA: Belknap Press of Harvard University Press.

Polanyi, Karl. 2001 [1944]. *The Great Transformation: The Political and Economic Origins of Our Time*. Boston, MA: Beacon Press.

Polanyi, Karl. 1957. "Aristotle Discovers the Economy." In *Trade and Market in the Early Empires*, edited by Arensberg, Polanyi, and Pearson. Chicago: Henry Regnery Company.

Portney, Paul, and Robert Stavins. 1998. "Market-based Environmental Policies." Discussion Paper 98–02. Kennedy School of Government, Harvard University. https://www.belfercenter.org/sites/default/files/legacy/files/disc_paper_98_02.pdf.

Poulantzas, Nicos. 1980 [1978]. *State, Power, Socialism*. London: Verso.

Powell, James Lawrence. 2011. *The Inquisition of Climate Science*. New York: Columbia University Press.

Przeworski, Adam. 1977. "Proletariat into a Class: The Process of Class Formation from Karl Kautsky's The Class Struggle to Recent Controversies." *Politics & Society* 7 (4). https://doi.org/10.1177/003232927700700401

Pulido, Laura. 1996. *Environmentalism and Economic Justice: Two Chicano Struggles in the Southwest*. Tucson: University of Arizona Press.

Pulido, Laura. 2015. "Geographies of Race and Ethnicity 1: White Supremacy vs White Privilege in Environmental Racism Research." *Progress in Human Geography* 39 (6): 809–817.

Pulido, Laura, Tianna Bruno, Cristina Faiver-Serna, and Cassandra Galentine. 2019. "Environmental Deregulation, Spectacular Racism, and White Nationalism in the Trump Era." *Annals of the American Association of Geographers* 109 (2): 520–532.

Purifoy, Danielle, and Louise Seamster. 2021. "Creative Extraction: Black Towns in White Space." *Society and Space D* 39 (1): 47–66.

Räthzel, Nora, Dimitris Stevis, and David Uzzell. 2021. *The Palgrave Handbook of Environmental Labor Studies*. London and New York: Palgrave Macmillan.

Rector, Josiah. 2014. "Environmental Justice at Work: The UAW, the War on Cancer, and the Right to Equal Protection from Toxic Hazards in Postwar America." *Journal of American History* 101 (2): 480–502.

Rector, Josiah. 2017. "Accumulating Risk: Environmental Justice and the History of Capitalism in Detroit, 1880–2015." PhD diss., Wayne State University.

Rector, Josiah. 2018. "The Spirit of Black Lake: Full Employment, Civil Rights, and the Forgotten Early History of Environmental Justice." *Modern American History* 1 (1): 45–66.

Rector, Josiah. 2022. *Toxic Debt: An Environmental Justice History of Detroit*. Chapel Hill: University of North Carolina Press.

Reed, Adolph. 1999. *The Underclass as Myth and Symbol. Stirrings in the Jug: Black Politics in the Post-Segregation Era*. Minneapolis: University of Minnesota Press.

Reed, Adolph. 2001. "Response to Eric Arnesen." *International Labor and Working-Class History* 60:69–80.

Reed, Adolph. 2013. "Marx, Race, and Neoliberalism." *New Labor Forum* 22 (1): 49–57.

Reed, Touré. 2008. *Not Alms but Opportunity: The Urban League and the Politics of Racial Uplift*. Chapel Hill: University of North Carolina Press.

Reed, Touré. 2020. *Toward Freedom: The Case against Race Reductionism*. London: Verso.

Rich, Nathaniel. 2019. *Losing Earth: A Recent History*. New York: Farrar, Straus and Giroux.

Rieder, Jonathan. 1989. "The Rise of the "Silent Majority"." In *The Rise and Fall of the New Deal Order*, edited by Steve Fraser and Gary Gerstle. Princeton, NJ: Princeton University Press.

Riley, Dylan. 2018. "What is Trump?" *New Left Review* 114. https://newleftreview.org/issues/ii114/articles/dylan-riley-what-is-trump

Roberts, J. Timmons, and Melissa Toffolon-Weiss. 2001. *Chronicles from the Environmental Justice Frontline*. Cambridge, UK: Cambridge University Press.

Robin, Corey. 2021. "Trump and the Trapped Country." *New Yorker*, March 13, 2021. https://www.newyorker.com/news/our-columnists/trump-and-the-trapped-country

Robinson, Cedric. 1983. *Black Marxism: The Making of the Black Radical Tradition*. Chapel Hill: University of North Carolina Press.

Roediger, David. 1991. *The Wages of Whiteness: Race and the Making of the American Working-Class*. London: Verso.

Romanello, Marina, Claudia Di Napoli, Paul Drummond, Carole Green, Harry Kennard, Pete Lampard, et al. 2022. "Countdown on Health and Climate Change: Health at the Mercy of Fossil Fuels." *The Lancet* 400 (10363): 1619–1654. https://doi.org/10.1016/S0140-6736(22)01540-9.

Rome, Adam. 2001. *The Bulldozer in the Countryside: Suburban Sprawl and the Rise of American Environmentalism*. Cambridge, UK: Cambridge University Press.

Ross, Benjamin, and Steven Amter. 2010. *The Polluters: The Making of Our Chemically Altered Environment*. Oxford: Oxford University Press.

Rothenberg, Randall. 1984. *The Neo-liberals: Creating the New American Politics*. New York: Simon and Schuster.

Rusher, William. 1975. *The Making of the New Majority Party*. Ottawa, IL: Green Hill Publishers.

Sale, Kirkpatrick. 1993. *The Green Revolution: The American Environmental Movement, 1962–1992*. New York: Hill and Wang.

Salleh, Ariel. 2003. "Ecofeminism as Sociology." *Capitalism, Nature, Socialism* 14 (1): 61–74.

Salleh, Ariel. 2005. "Moving to an Embodied Materialism." *Capitalism, Nature, Socialism* 16 (2): 9–14.

Saxton, Alexander. 1990. *The Rise and Fall of the White Republic: Class Politics and Mass Culture in Nineteenth-Century America*. London and New York: Verso.

Schnaiberg, Allan. 1980. *The Environment: From Surplus to Scarcity*. New York: Oxford University Press.

Schnaiberg, Allan, David Pellow, and Adam Weinberg. 2002. "The Treadmill of Production and the Environmental State." In *Research in Social Problems and Public Policy*, edited by Mol and Buttel, 15–32.. Leeds, UK: Emerald Publishing.

Sellers, Christopher. 1997. *Hazards of the Job: From Industrial Disease to Environmental Health Science*. Chapel Hill, NC: University of North Carolina Press.

Sellers, Christopher. 2012. *Crabgrass Crucible: Suburban Nature and the Rise of Environmentalism in Twentieth-Century America*. Chapel Hill: University of North Carolina Press.

Silva, Jennifer. 2019. *We're Still Here: Pain and Politics in the Heart of America*. Oxford: Oxford University Press.

Silver, Beverly. 2003. *Forces of Labor: Workers' Movements and Globalization since 1870*. Cambridge, UK: Cambridge University Press.

Silver, Nate. 2016. "The Mythology of Trump's Working-Class Support." *FiveThirtyEight*, May 3. https://fivethirtyeight.com/features/the-mythology-of-trumps-working-class-support/.

Skocpol, Theda. 2013. "Naming the Problem: What It Will Take to Counter Extremism and Combat Global Marming." Prepared for the Symposium on The Politics of America's Fight against Global Warming, Harvard University, February 14.

Slotkin, Richard. 1998 [1985]. *The Fatal Environment: The Myth of the Frontier in the Age of Industrialization, 1800–1890*. Norman: University of Oklahoma Press.

Smith, J. Douglass. 1998. "'When Reason Collides with Prejudice': Armistead Lloyd Boothe and the Politics of Moderation." In *The Moderates' Dilemma: Massive Resistance to School Desegregation in Virginia*, edited by Lassiter and Lewis, 22–50. Charlottesville: University of Virginia Press.

Smith, Jessica. 2017. "Blind Spots of Liberal Righteousness." https://culanth.org/fieldsights/blind-spots-of-liberal-righteousness.

Smith, Kimberly. 2007. *African American Environmental Thought: Foundations*. Lawrence: University Press of Kansas.

Spence, Mark David. 1999. *Dispossessing the Wilderness: Indian Removal and the Making of the National Parks*. Oxford, UK: Oxford University Press.

Sugrue, Thomas. 1996. *The Origins of the Urban Crisis: Race and Inequality in Postwar Detroit*. Princeton, NJ: Princeton University Press.

Stein, Judith. 1998. *Running Steel, Running America: Race, Economic Policy and the Decline of Liberalism*. Chapel Hill: University of North Carolina Press.

Stein, Judith. 2010. *Pivotal Decade: How the United States Trade Factories for Finance in the Seventies*. London: Yale University Press.

Stein, Rachel, ed. 2004. *New Perspectives on Environmental Justice: Gender, Sexuality, and Activism*. New Brunswick, NJ: Rutgers University Press.

Stevis, Dimitris. 2011. "Unions and the Environment: Pathways to Global Labor Environmentalism." *WorkingUSA* 14:145–159

Stevis, Dimitris. 2023. *Just Transitions: Promise and Contestation*. Cambridge, UK: Cambridge University Press.

Stevis, Dimitris, and Romain Felli. 2016. "Green Transitions, Just Transitions? Broadening and Deepening Justice." *Kurswechsel* 3:35–45.

Strange, Susan. 1986. *Casino Capitalism*. Manchester, UK: Manchester University Press.

Sunshine, Spencer. "Smokescreen: How Timber Unity Has Mainstreamed Militia Groups, Alt Right, and Conspiracy Theories in Oregon Politics." Self-published. https://spencersunshine.com/2020/03/04/smokescreen/.

Switzer, Jacqueline Vaughn. 1997. *Green Backlash: The History and Politics of Environmental Opposition in the U.S.* Boulder, CO: Lynne Rienner.

Táíwò, Olúfẹ́mi. 2022. *Elite Capture: How the Powerful Took Over Identity Politics (and Everything Else)*. Chicago: Haymarket Books.

Taylor, Dorceta. 2014. *Toxic Communities: Environmental Racism, Industrial Pollution, and Residential Mobility*. New York: New York University Press.

Taylor, Dorceta. 2016. *The Rise of the American Conservation Movement: Power, Privilege, and Environmental Protection*. Durham, NC: Duke University Press.

Taylor, Keeanga-Yamahtta. 2019. *Race for Profit: How Banks and the Real Estate Industry Undermined Black Homeownership*. Chapel Hill: University of North Carolina Press.

Teixeira, Ruy. 2021. "Ten Things We Now Know about the Nonwhite Working-Class Vote in 2020. *Liberal Patriot*, July 1, 2021. https://www.liberalpatriot.com/p/ten-things-we-now-know -about-the-62a.

Teixeira, Ruy. 2022. "The Democrats' Working-Class Voter Problem." *Liberal Patriot*, March 10, 2022. https://www.liberalpatriot.com/p/the-democrats-working-class-voter.

Thompson, E. P. 1963. *The Making of the English Working-Class*. New York: Vintage Books.

Thompson, Jonathan. 2016. "The First Sagebrush Rebellion: What Sparked It and How It Ended." *High Country News*, January 14, 2016. https://www.hcn.org/articles/a-look-back-at-the-first -sagebrush-rebellion.

Tingling, Michele. 1980. "Cities: The Urban Environment." *Environment* 22 (3): 5–42.

Tocqueville, Alexis. 2003. *Democracy in America*. Washington, DC: Regnery Publishing.

Trent, Mark, and Campbell Robertson. 2018. "Despair, Love and Loss: A Journey inside West Virginia's Opioid Crisis." *New York Times*, December 13, 2018. https://www.nytimes.com /interactive/2018/us/west-virginia-opioids.html.

Tucker, William. 1982. *Progress and Privilege: America in the Age of Environmentalism*. Garden City, NY: Anchor Press/Doubleday.

Turner, James Morton. 2009. "'The Spectre of Environmentalism': Wilderness, Environmental Politics, and the Evolution of the New Right." *Journal of American History* 96 (1): 123–148.

Turner, James Morton, and Andrew Isenberg. 2018. *The Republican Reversal: Conservatives and the Environment from Nixon to Trump*. Cambridge, MA: Harvard University Press.

Udall, Stuart. 1963. *The Quiet Crisis*. New York : Avon Books.

United Church of Christ. 1987. "Toxic Wastes and Race in the United States." Commission on Racial Justice. https://www.ucc.org/wp-content/uploads/2020/12/ToxicWastesRace.pdf.

Upin, Catherine. 2012. "Climate of Doubt." PBS. *Frontline*, October 23, 2012.https://www.pbs .org/wgbh/frontline/documentary/climate-of-doubt/

Vogel, David. 1983. "The Power of Business in America: A Re-Appraisal." *British Journal of Political Science* 13 (1): 19–43.

Voyles, Traci Brynne. 2015. *Wastelanding: Legacies of Uranium Mining in Navajo Country*. Minneapolis: University of Minnesota Press.

Wacquant, Loïc. 2010. "Class, Race, and Hyperincarceration in Revanchist America." *Daedalus* 139 (3): 74–90.

Wacquant, Loïc. 2022. *The Invention of the Underclass: A Study in the Politics of Knowledge*. Cambridge, UK: Polity.

Walter, Amy. 2021. "Democrats Lost Ground with Non-College Voters of Color in 2020." *Cook Political Report,* June 17, 2021. https://www.cookpolitical.com/analysis/national/national-politics/democrats-lost-ground-non-college-voters-color-2020

Ward, Brandon. 2019. "Suburbs against the Region: Homeowner Environmentalism in 1970s Detroit." *Journal of Planning History* 18 (2): 83–101.

Washington, Sylvia Hood. 2005. *Packing Them In: An Archaeology of Environmental Racism in Chicago, 1865–1954.* Lanham, MD: Lexington Books.

Weber, Max. 1919. "Politics as a Vocation." http://fs2.american.edu/dfagel/www/class%20readings/weber/politicsasavocation.pdf.

Wells, Christopher. 2012. *Car Country: An Environmental History.* Seattle: University of Washington Press.

West, Cornel. 2016. "Goodbye, American Neoliberalism. A New Era Is Here." *The Guardian,* November 17, 2016. https://www.theguardian.com/commentisfree/2016/nov/17/american-neoliberalism-cornel-west-2016-election.

White, Deborah Gray. 1985. *Ar'n't I a Woman? Female Slaves in the Plantation South.* New York: W.W. Norton and Company.

White, Lynn, Jr. 1967. "The Historical Roots of Our Ecologic Crisis." *Science* 155 (3767): 1203–1207.

White, Richard. 1991. *It's Your Misfortune and None of My Own: A New History of the American West.* Norman: University of Oklahoma Press.

White, Richard. 1995. *The Organic Machine: The Remaking of the Columbia River.* New York: Hill and Wang.

White, Richard. 1996. "Are You an Environmentalist or Do You Work for a Living?': Work and Nature." In *Uncommon Ground: Rethinking the Human Place in Nature,* edited by William Cronon.. New York: WW Norton & Company.

Whyte, Kyle. 2016. "Indigenous Environmental Movements and the Function of Governance Institutions." In *The Oxford Handbook of Environmental Political Theory,* edited by Teena Gabrielson, Cheryl Hall, John Meyer and David Schlosberg. Oxford, UK: Oxford University Press.

Whyte, William. 1956. *The Organization Man.* New York: Simon and Schuster.

Wilkerson, Isabel. 2010. *The Warmth of Other Suns: The Epic Story of America's Great Migration.* New York: Random House.

Williams, Raymond. 2005 [1972]. "Ideas of Nature." In *Nature: Critical Concepts in the Social Sciences,* edited by David Inglis, John Bone, Rhoda Wilkie, 47–62. New York: Routledge.

Williams, Trina. 2000. "The Homestead Act: A Major Asset-Building Policy in American History." Washington University-St. Louis, Center for Social Development, Working Paper 00–9.

Wolin, Sheldon. 1996. "Fugitive Democracy." In *Democracy and Difference: Contesting the Boundaries of the Political,* edited by Seyla Benhabib, Princeton, NJ: Princeton University Press.

Wood, Robert C. 1958. *Suburbia: It's People and Their Politics.* Boston: Houghton Mifflin.

Wright, Eric Olin. 1980. "Varieties of Marxist Conceptions of Class Structure." *Politics and Society* 9 (3): 323–370.

Zetkin, Clara. 1923. "The Struggle Against Fascism." June 20, 1923. https://www.marxists.org /archive/zetkin/1923/06/struggle-against-fascism.html.

Zimring, Carl. 2017. *Clean and White: A History of Environmental Racism in the United States.* New York: New York University Press.

Zweig, Michael. 2012. *The Working-Class Majority: America's Best Kept Secret.* Ithaca, NY: Cornell University Press.

Index

Calhounian doctrine of interposition, 74
California
    suburban environment in, 72, 75–76
    Yosemite land grant in, 53
Cannon, Lou, 107
Cap-and-trade systems, 132, 134, 135–136
    Green New Deal compared to, 170
    New Democrats on, 125
    opposition of Manchin to, 179
    Oregon bill on, 143–144, 145, 186
Capital
    fossil (*see* Fossil capital)
    human, 148–150
    in neoliberalism, 149
    woke, 183
Capital accumulation, 158
    in class-ecological orders, 18
    force used in, 13–14
    in fossil fuel use, 14–15
    natural resources for, 15, 53, 184
    for owners of means of production, 13
    and racism, 8
    social reproductive labor required for, 8, 9,
        15, 64, 169, 210n6
    state role in, 9, 10–11
    in treadmill of production, 61, 84
Capitalism, 13, 15, 19
    exploitation in, 9, 15
    fossil, 14–15
    global, 18, 137, 197
    green, 16, 200
    industrial (*see* Industrial capitalism)
    racial, 67
    social reproductive labor in, 64, 168
    state power in, 10
Capitalist class (ruling class), 10–12
    anti-environmentalism in, 23, 25, 91,
        92–94, 173
    cheap natures required by, 15
    fossil capital in, 16–17 (*See also* Fossil
        capital)
    fractions of, 6

in frontier era, 38, 39
global, 137
in industrial production, 38
investments in smoke and spoil, 16
in nature preservation movements, 16
and ruling class ecology, 50–52, 53
and ruling class-for-itself, 6, 18, 107–113,
    115
and ruling class-in-itself, 5, 6, 16, 17–18
and state relations, 10–11
in treadmill of production, 24, 59, 60,
    61–62, 109, 110
and underclass, 119
whiteness in, 31–33
and working class relations, 5, 7, 8, 10–12,
    15, 31–32, 38, 39, 52, 111–115
Capitol insurrection, 17, 144, 192–193
Carbon dioxide emissions, 1, 98, 131
Carbon monoxide, 1
Carbon tax, 132, 133
Carceral state, 23, 119, 120, 126, 140, 191
Care work
    of indigenous peoples, 171
    and social reproduction, 8–9, 57, 64–67,
        169–170, 189
    as unwaged labor of women, 37, 64, 65
Carson, Rachel, 81, 214n7
Carter, Jimmy, 100, 106
Catholic Irish immigrants, 33, 45, 46
Cato Institute, 108
Cattle barons in frontier era, 48, 51
Cattlemen's Association, 101, 127
Center for Coal-field Justice, 164
Center for the Defense of Free Enterprise, 102
Central American immigrants, 4, 202
Central Pacific Railroad, 48
Chamber of Commerce, 6, 84, 92, 94, 108,
    127, 132, 133
Cheap natures, 15
Chemicals
    in agriculture, 14, 61, 85, 89, 213n23
    bioaccumulation of, 25, 61

Democratic Party (cont.)
in frontier era, 44
and Green New Deal, 178–179, 198, 200
and Irish immigrants, 45
and neoliberalism, 190–191
and New Democrats, 121–126 (see also
New Democrats)
as pro-slavery, 33, 45, 184
in suburbs, 90–91, 137, 213n26
working class in, 44, 121, 181–182
Democratic Socialism, 198, 199–200
and Green New Deal, 178, 194
Demos, 24, 148, 150, 168, 176
Developing countries, Kyoto Protocol on,
133–134
DiAngelo, Robin, 31
Dictatorship of capital, 51
Dictatorship of labor, 50
Dimon, Jamie, 180
Discrimination, 187–188, 209n3
in employment, 85, 91, 120, 212n20
energy, 182
in housing (see Housing discrimination and
segregation)
Dissent (Kuttner), 111
Diversity, equity, and inclusion (DEI) train-
ing, 31, 188
Djilas, Milovan, 115
Donohue, Thomas, 133
Double movement or counter-movement,
147, 190
Douglass, Frederick, 45
Drug addiction and overdose, 24, 88, 153,
160, 161, 162
Du Bois, W. E. B., 21–22, 28, 37
on abolition-democracy, 51
Black Reconstruction in America, 21–22, 28,
30, 35, 208–209n1
on cotton production, 38, 46
on demands of Black workers, 50
on enslaved people and abolitionist allies, 41
on forgotten mass of men, 46–47

on Freedmen's Bureau, 50
on immigrant wage workers, 43, 44
on land speculation in West, 48
on natural resources of South, 46
on racial divisions in working class, 184
on wages of whiteness, 21–22, 28, 29–30,
52, 186
and whiteness studies, 21, 28, 30, 31
on white workers in South, 46
Dunlap, Riley E., 186
DuPont, 17, 132, 135

Earth Day, 91, 92, 94
Earth First!, 128, 129
Eastland, James, 74
Ebell, Myron, 136
Eco-capitalists, 16
Eco-fascism, 161
Eco-feminists, 66, 159
Ecological revolutions, Merchant on, 12–13,
65, 207–208n4
Ecology, 13, 65
origin of term, 156, 158, 216n6
Eco-nationalism, 195
Economic Dislocation Early Warning System
of EPA, 112
Economy
Aristotelian view of, 157
as autonomous sphere of life, 158
confidence in, 10–11
extractive, in Trump country, 21
in industrial capitalism, 158
inequality in, 138
informal, 9, 119
knowledge, 137
origin of term, 156, 172
plantation (see Plantation economy)
post-carbon, 200
Eco-socialism, 172, 197–200
and just transition, 203, 204, 205
on production and social production, 170,
201

on relationship between class and environment, 19, 20
on whiteness, 32
Ecosystem withdrawals and additions for
production, 60–61, 63, 94–95, 184,
189
Ecuador, 17, 139
Education
of formerly enslaved, 50
local choice of schools in, 56, 73, 210n2
and political party affiliation, 198, 199
school segregation in, 56, 210nn2–4
in Trump country, 153, 155, 215n4
in "University of Reclamation," 201
and wages, 186
Ehrenreich, Barbara, 115, 117, 121
Ehrlich, Paul and Anne, 88
Elections
first-past-the-post, 11, 136
gerrymandering of districts in, 11–12, 192
and presidential campaign (1964), 76–77
and presidential campaign (2016),
151–152, 215n2
single-member districts in, 11, 136
Electoral college, 11
Electrical vehicles, 204, 205
Elites, economic. *See* Capitalist class (ruling
class)
Elitists, environmental, 28
and climate change policies, 5, 136
and Earth Day, 92
and extractive laborers, 4–5, 23, 128
and New Class, 23, 102, 103, 115–117,
185
Employment
automation affecting, 60, 64, 84
in automobile factories, 60, 85, 204–205,
212n19, 215n5
of Black women, unequal pay in, 210n7
of Black workers in frontier era, 40–41
in blue-collar jobs, 84 (*see also* Blue-collar
workers)

deindustrialization affecting, 7, 110
discrimination in, 85, 91, 120,
212n20
environmental laws as threat to (*See* Jobs
versus environment)
of immigrants, 33, 68
in Shell Pennsylvania Petrochemicals
Complex, 1–2
in treadmill of production, 60, 62–64
wages in (*see* Wages)
in white-collar jobs, 84, 97
whiteness affecting, 30, 33
of women, 65, 104, 138, 210nn7–8
workplace hazards in, 41, 85, 89, 91,
213n23
Enbridge pipelines, 167, 174, 175
Enclosure or fence laws, 52, 70–71, 160, 185,
209n3
Endangered Species Act, 59, 127, 128
amendments to, 107
Energy discrimination, 182
Energy Independence executive order, 28
Energy use
alternative sources in, 16, 17, 99, 102
nuclear power in, 99, 100, 102
and oil crisis, 105–106
in treadmill of production, 22, 57, 60, 61,
94, 189
Engels, Friedrich, 10, 14, 64
Enslaved peoples, 35, 40–42
and abolitionist allies, 29
in cotton production, 15, 38
general strike of, 49
and land distribution to formerly enslaved,
49–50, 51, 209n4
in three-fifths compromise, 46
unfreedom of, 37
and Union Army enlistment of formerly
enslaved, 49, 50
and wilderness, 41–42
women as, 42–43
Enterprise system, 108

Entrepreneurship, New Democrat support
of, 123
Environment, concept of, 70, 81
Environmental, social, governance (ESG)
approach to investing, 182, 217n1
Environmental Defense Fund, 81, 123, 124,
135, 213n24
Environmental harm/degradation
in air pollution (*see* Air pollution)
cap-and-trade policies on (*see* Cap-and-
trade systems)
from climate change events, 160, 191,
194–195
from coal industry, 3, 162–164, 172
conservation policies on, 53, 55 (*see also*
Conservationism)
disaster fund for, 213n25
from fossil fuels, 1, 3, 15, 17, 98, 131, 136
global communities exposed to, 139
immigrants exposed to, 68
inequality in exposure to, 63, 89, 103,
138–142, 184, 187
Native Americans exposed to, 139,
140–141, 165
in noise pollution, 83
permits for, 124, 125
in post–World War II era, 57
as price to pay for jobs, 154–155, 174, 202
(*see also* Jobs versus environment)
from production processes, 57, 59–64,
94–95, 184
racial response to, 34, 212n21
residential mobility in avoidance of, 89,
187
from Shell Pennsylvania Petrochemicals
Complex, 1, 3
in soil contamination, 21, 79, 139, 191
in suburbs, 78–83, 211n15
as threat to home, 15, 159, 163, 165, 191
from timber industry, 128, 129
in Trump country, 146
and "University of Reclamation," 201

in urban areas, 87, 88, 89, 139–140,
213n25
in water pollution (*see* Water pollution)
whiteness reducing exposure to, 140, 184
working class affected by, 15, 63, 140–141,
175, 178, 181–182, 187, 202
working class resistance to, 144–145
in workplace, 41, 85, 91
Environmentalism
activism in (*see* Activism)
climate change concerns in, 131–137
elitists in (*see* Elitists, environmental)
forest preservation concerns in, 127–129
hyper-politicization of, 115
as job threat, 23 (*see also* Jobs versus environment)
and labor movements, 92, 168–171
and labor unions, 91, 102, 117, 130–131,
168, 169, 213n27
laws and regulations in (*see* Laws and
regulations, environmental)
leisure concerns in, 91
market-based, 107
in middle class, 28, 91, 92
and nature, 81, 88, 211n14
and New Democrats, 22, 123–125
and oil crisis, 105, 106
in suburbs, 22–23, 58, 78–83, 88–89,
91–94, 96, 97, 190
and treadmill of production, 60, 62, 63, 64
in Virginia, 55, 82, 95–96, 209n1
Wise Use opposition to, 126–130
in working class, 28, 102, 144–145, 188
workplace concerns in, 91
Environmentalists for Full Employment
(EFFE), 99, 102, 107, 112, 124, 139,
203
Environmental justice, 92, 103, 138–142,
169, 191
and home as socio-ecological site, 159
job concerns in, 168
opposition to, 183

Northern areas of US, 35
abolition-democracy and ruling class ecology in, 50, 51
dictatorship of capital in, 51
dictatorship of labor in, 50
migration of Blacks to, 68–69
in three-fifths compromise, 44, 46
Northern Pacific Railroad, 48
Northern spotted owls, 128
North Star, The, 41, 209n2
Nuclear power, 99, 100
and Three Mile Island accident, 100, 102, 160
Nunes, Devin, 135

Obach, Brian, 130–131
Obama, Barack, 134–135, 154, 195
Ocasio-Cortez, Alexandria, 178
Ocasio-Cortez-Markey resolution for Green New Deal, 170, 178
Occupational Safety and Health Administration, 91, 112
Occupy Wall Street, 147, 197
O'Connell, Daniel, 45
Ohio, 111, 153, 191
Oikonomia, 157
Oikos, 24, 156–161, 165, 167, 168
environmental threats to, 175
as socio-ecological site, 24, 159–162, 176, 191, 195
traditional, conservatives on threats to, 171–172, 216n11
and undervalued labor in social reproduction, 170
Oil, Chemical, and Atomic Workers (OCAW), 91, 111–112
in Louisiana, 141, 169
on Superfund for Workers, 200
on Worker's Bill of Rights, 201
Oil and gas industry
automation and job loss in, 60
during Bush administration, 134

climate change mitigation as threat to, 17, 131, 194
cost of crude oil in, 105
in energy crisis, 105–106
Inflation Reduction Act benefits for, 179–180
on Native American lands, 139
primary fossil capital in, 17
production trends in, 61
profits in, 4
and Sagebrush Rebellion, 101
Shell Pennsylvania Petrochemicals Complex in, 1–3
subsidies for, 62
in treadmill of production, 61, 62
during Trump administration, 28
water pollution from, 154
Wise Use support of, 127
Olszanski, Mike, 102
Opioid addiction and overdose, 160, 161
Oregon
cap-and-trade bill in, 143–144, 145, 186
indigenous activists in, 167
timber industry in, 128, 143–144
Oregon Lands Coalition, 128
Origin of the Family, Private Property and the State (Engels), 64
Origins of Totalitarianism, The (Arendt), 195
Others, 24, 58, 77, 195
Overton window, 198
Owls, spotted, 128, 129, 130
Owners of means of production, 13, 37
Ozone, 3, 131

Pacific Northwest timber industry, 127–129, 208n5
Painter, Nell Irvin, 45, 74
Palingenetic ultranationalism, 193
Particulate matter production, 1, 3
Patel, Raj, 12, 15
Patriarchy, social reproductive labor in, 64
PCB contamination of soil, 139